THE FLORIDA MEADOW MANUAL

UNIVERSITY PRESS OF FLORIDA
Florida A&M University, Tallahassee
Florida Atlantic University, Boca Raton
Florida Gulf Coast University, Ft. Myers
Florida International University, Miami
Florida State University, Tallahassee
New College of Florida, Sarasota
University of Central Florida, Orlando
University of Florida, Gainesville
University of North Florida, Jacksonville
University of South Florida, Tampa
University of West Florida, Pensacola

Gage Daniel J. LaPierre and Isabella Guttuso Browne

the FLORIDA MEADOW MANUAL

Creating *and* Maintaining Resilient Green Spaces

UNIVERSITY PRESS OF FLORIDA
Gainesville / Tallahassee / Tampa / Boca Raton
Pensacola / Orlando / Miami / Jacksonville / Ft. Myers / Sarasota

Cover: Photo by NaMa Native Landscapes. *Spine*: Photo by Lilly B. Anderson-Messec. *Frontis*: Photo by Isabella Guttuso Browne. *Text and cover design*: Mindy Basinger Hill

Published in the United States of America
Design by Mindy Basinger Hill / Composed in Calluna Pro and Calluna Sans Pro

31 30 29 28 27 26 6 5 4 3 2 1

LIBRARY OF CONGRESS CATALOGING-IN-PUBLICATION DATA

Names: LaPierre, Gage Daniel J. author | Guttuso Browne, Isabella author

Title: The Florida meadow manual : creating and maintaining resilient green spaces / Gage Daniel J. LaPierre, Isabella Guttuso Browne.

Description: Gainesville : University Press of Florida, 2026. | Includes bibliographical references and index.

Identifiers: LCCN 2025053407 (print) | LCCN 2025053408 (ebook) | ISBN 9780813081557 paperback | ISBN 9780813074313 pdf | ISBN 9780813075426 ebook

Subjects: LCSH: Meadow gardening—Florida | Meadow plants—Florida | Grassland plants—Florida | Landscape design—Florida | BISAC: GARDENING / Landscape | SCIENCE / Life Sciences / Horticulture | LCGFT: Handbooks and manuals

Classification: LCC SB439.24.F6 L37 2026 (print) | LCC SB439.24.F6 (ebook)

LC record available at https://lccn.loc.gov/2025053407

LC ebook record available at https://lccn.loc.gov/2025053408

The University Press of Florida is the scholarly publishing agency for the State University System of Florida, comprising Florida A&M University, Florida Atlantic University, Florida Gulf Coast University, Florida International University, Florida State University, New College of Florida, University of Central Florida, University of Florida, University of North Florida, University of South Florida, and University of West Florida.

University Press of Florida
PO Box 140239
Gainesville, FL 32614
floridapress.org

GPSR EU Authorized Representative: Mare Nostrum Group B.V., Mauritskade 21D, 1091 GC Amsterdam, The Netherlands, gpsr@mare-nostrum.co.uk

CONTENTS

PREFACE

Picture a sunlit scene—a common occurrence in Florida—but instead of the expected uniformity of a mowed lawn or the structured formality of a designed garden, envision a dynamic, living mosaic unfolding before you. Here, native grasses sway amid bursts of color from wildflowers, and the air thrums with the movement of pollinators, birds, and the wind. This is the essence of a Florida meadow—a dynamic, intentionally crafted plant community that honors and embraces wild beauty while inviting human stewardship. More than just an arrangement of pretty flowers, a Florida meadow, as explored within these pages, represents a conscious partnership with nature, a deliberate act of ecological enhancement tailored to the unique conditions of the Sunshine State.

The Florida Meadow Manual is your invitation to cultivate such places, whether on a grand scale or within the intimate confines of a backyard. Rooted in Florida's rich ecological heritage and shaped by the pressing need to support biodiversity in an era of increasing habitat destruction, this book offers both inspiration and instruction. It reflects a growing movement that seeks not just to conserve what's left, but to create anew: resilient, native-rich landscapes where all life can thrive. We believe that through the knowledge compiled in this book, individuals and communities across Florida can transform underutilized or ecologically impoverished lands into thriving havens for native plants and wildlife.

The journey to compile *The Florida Meadow Manual* has been one of deep exploration and synthesis, drawing upon a rich and varied wellspring of knowledge. At its heart lies the accumulated wisdom of many professionals who have dedicated their lives to understanding and working with Florida's

opposite A meadow. Photo by Lilly B. Anderson-Messec.

native plants and plant communities. Among them are seasoned horticulturalists, botanists, conservationists, landscape architects, ecologists, and dedicated researchers. Their practical insights, often gleaned through years of meticulous observation, experimentation, and an intimate connection with Florida's diverse flora, fauna, and ecological processes, form an invaluable cornerstone of this book. Their combined experience and knowledge, enriched by both successes and the lessons learned from challenges in restoring and re-creating Florida's spectacular biological diversity, are woven throughout the chapters, offering a tangible and relatable pathway to successful meadow creation.

This text is not just a how-to; it is a why-to. By marrying experiential knowledge with credible research, we aim to offer a comprehensive and accessible guide. Each chapter is designed to help you navigate the process of meadow creation, from understanding your site's ecological potential to selecting appropriate native species, preparing the site, and implementing planting and management strategies. Whether you are sowing seed or installing plugs, managing weeds, or planning for prescribed fire, the guidance herein is tailored to Florida's unique setting.

Ultimately, this book is about more than creating meadows. It's about participating in something larger: a quiet, hopeful act of caring for the land that binds us together. To build a Florida meadow is to co-create with nature, to work with patience and humility, and to witness how even small patches of land can pulse with life and purpose. Ultimately, we believe that Florida meadows offer a powerful opportunity to reconnect with the natural world through creating beautiful and functional spaces that enrich our lives and contribute to a healthier environment for generations to come.

As you turn these pages, you'll find guidance, yes, but also a mindset. We encourage you to observe, to adapt, and to embrace the long view. Creating a meadow is not a one-time project but an evolving partnership with place. We hope this manual helps you find your footing on that journey and empowers you to nurture your own corner of Florida's wild heart.

Special Thanks

This book would not have been possible without knowledge collected from countless researchers and many working professionals. The following people were invaluable sources of firsthand experiences and knowledge: Paul Abel,

Nancy Bissett, Edwin Bridges, Chris Matson, Joe Reems, and Troy Springer. M. J. "Mimi" Williams also provided invaluable assistance through her thorough review of the text, as well as her expertise on seed mixture selection, site preparation techniques, and ongoing management strategies.

This book would not have been possible without great encouragement from Cammie Donaldson and funding from the Native Plant Horticultural Foundation, the Florida Association of Native Nurseries, and the Florida Wildflower Foundation.

THE FLORIDA MEADOW MANUAL

one

THE MEADOW CONCEPT

Hope is a walk through a flowering meadow.
One does not require that it lead anywhere.
ROBERT BREAULT

Meadows present an exciting evolution in how we envision urban green spaces and degraded lands. We begin by looking at how current landscaping practices, such as lawns, became such a fixture in the American mind. Then we introduce the concept of a meadow as an intentionally designed and maintained novel plant community. Finally, we spend the majority of the chapter describing the benefits of creating meadows, ranging from aesthetic to environmental to educational. We trust that by the end of these pages, you'll share our enthusiasm for the sustainable future that meadows open up in Florida and throughout the country.

The American Lawn Cultural Paradigm

Lawns arguably have played a significant role in American history and culture. Indeed, lawns are viewed as quintessentially American. However, the rise of the American lawn did not happen overnight. It was a gradual process, spanning more than two centuries and driven by a combination of government initiatives, industrial marketing, and societal trends. These included government support of suburban developments that mandated lawns; the mass production of affordable lawn equipment; widespread advertising campaigns that promoted lush, green lawns as a symbol of prosperity and social standing; and, crucially, the implementation of local ordinances requiring individuals to maintain lawns to certain standards. These ordinances often dictated specific lawn heights, prohibited "weeds," and enforced conformity

opposite Photo by Shelby Radcliffe with Emergent Gardens.

within neighborhoods, solidifying the lawn as a mandatory feature of residential landscapes. These efforts successfully tied the image of a pristine, well-maintained lawn to notions of career success, community belonging, family values, and the ideal suburban lifestyle.[1]

The symbolic value of lawns arose partly because lawns take money and energy to create and maintain, acting as a visible display of upward mobility and adherence to middle-class values. This perception has roots in lawns' original emergence in the higher aristocracy of Europe, where vast, manicured estates signaled wealth and status. In addition, lawns became the norm in many of the first planned suburban developments after the Civil War, a trend that accelerated after both World Wars.[2]

The cultural pressure to maintain a "good" lawn became deeply ingrained in US society, driven by a confluence of factors. This peer pressure manifested in neighborhood norms, social expectations, and increasingly, the rise of homeowners' associations (HOAS). HOAS, in particular, played a significant role in enforcing conformity, often through strict regulations on lawn height, weed control, and overall appearance, reinforcing the idea of a uniform, monoculture turfgrass lawn as the ideal.

The evolution of lawn care also saw a shift from labor-intensive, high-investment practices to more accessible, though still demanding, maintenance. Early lawns required substantial manual labor and often involved hiring gardeners. Later, the development of motorized lawnmowers, chemical fertilizers, and pesticides democratized lawn care, making it achievable for a wider range of homeowners. Yet, despite the reduced labor, the cultural expectation of a perfectly manicured lawn persisted, even intensified, as it became a symbol of community belonging, social class, and personal responsibility. Advertisements and media representations further solidified this ideal, portraying the lush, green lawn as an integral part of the American dream.

As a result, lawns currently blanket 40 million acres of land in the United States.[3] This is roughly 2 percent of the continental United States, or the size of the entire state of Florida. Such prevalence has produced a multi-billion-dollar industry centered on lawn care.[4] At this scale lawns have become a major contributor to many environmental issues, ranging from climate change to the biodiversity crisis, eutrophication of waterways, pollinator population collapse, and even health issues. Today, lawn care research is attempting to mitigate some of these issues with improved varieties of turfgrass, newer technology, and smarter maintenance practices. Fundamentally, however, lawns

One of the first preplanned, mass-produced uniform suburban communities was Levittown, Long Island, New York. The uniform lawns quickly became the standard for all suburbs. New York State Archives, aerial photographic prints and negatives of New York state sites, October 7, 1947, NYSA_B1598-99_102399.

Landscaping for all the homes in Levittown, Long Island, New York, featured uniform, closely mowed monoculture lawns. Gottscho-Schleisner Collection, Library of Congress, Prints & Photographs Division, LC-G613-72793.

in the United States are unsustainable at this scale, rendering these solutions unlikely to offer significant hope in curbing ongoing environmental issues.

As a result, many are calling for a reinvention of the American lawn into something that supports biodiversity, sustainability, and healthy communities. This does not necessarily entail complete removal of lawns but could be simply a reduction of lawn size in favor of native landscaping or promoting mixed polyculture native lawns. One significant opportunity for the reinvention of the American lawn is to incorporate meadows as a supporting, functional element (for example, pathways, edging, recreational space). Such measures may help preserve the positive functional elements of the American lawn while reintroducing Americans to indigenous American flora and nature.

What Are Meadows?

For many, the word **meadow** evokes images of a picturesque, open grassy area dotted with wildflowers. Indeed, most dictionaries align with this perception, offering similarly broad definitions. However, this generalization has led the term to be applied to a wide range of plant communities, both naturally occurring systems (such as alpine meadows at high elevations and wet

A typical suburban Florida home with a St. Augustine turfgrass lawn. Photo courtesy of https://www.pickpik.com.

left A large meadow installed in a central Florida yard, completely replacing a lawn. Photo by Harriet Festing.

right A residential yard that retains some lawn for recreational use with a meadow around the periphery. Photo by Heather Crane, Native Plant Horticulture Foundation.

meadows in perennially moist areas) as well as human-created landscapes. Technically speaking, no natural communities in Florida are classified as meadows.[5] Therefore, in this text we define a **meadow** in Florida as any novel plant community designed or maintained by humans and dominated by a diversity of grasses and sedges (**graminoids**), as well as herbaceous flowering plants (**forbs**). Unlike gardens, which are usually meticulously structured, intensively managed, and centered on ornamental plants, meadows are dynamic, seminatural systems that emphasize ecological function and a wilder aesthetic. Ultimately, meadows are not merely grassy fields or wildflower stands but rather purposefully created plant communities or ecosystems with intentional composition designed for ecological benefit.

Meadows Versus Savannas, Prairies, and Grasslands

As defined, **meadow** is a broad term that shares a great deal of overlap with prairies, grasslands, and savannas.[6] These other terms often have distinct uses in different regions. **Grassland**, for instance, is an umbrella term often applied to differentiate groups of ecosystems from forested ecosystems. Likewise, **savanna** is also an umbrella term for grassland ecosystems that feature moderate tree cover but not to the level of a forest. In contrast, **prairie**

Pine savanna at Austin Cary Forest. Photo by Gage LaPierre.

A wet prairie in Highlands County,Florida. In wet prairies, the soil is moist or even saturated most of the time. Photo by Gage LaPierre.

Dry prairie at DeLuca Ranch, Florida. Photo by Gage LaPierre.

Seepage slope prairie community in Apalachicola National Forest, Florida, featuring yellow pitcher plants. Photo by Gage LaPierre.

usually refers to more localized or specific types of natural communities (such as Payne's Prairie, Kissimmee River Prairie, Wet Prairie community).[7]

What Is a Native Plant?

A **native plant** is generally defined as a species that has evolved and occurred naturally in a specific region over a long period, typically before significant human intervention such as European colonization. These plants have adapted to the local climate, soils, and ecological interactions with other native species, including pollinators, herbivores, and microbes. Because of these adaptations, native plants play essential roles in maintaining ecosystem stability, supporting biodiversity, and contributing to natural processes such as nutrient cycling and water filtration. However, defining what is truly "native" can be complex, because plant distributions have shifted over millennia due to climate changes, animal dispersal, and more recently, human influence.

One major issue in defining native plants is determining the appropriate geographic and temporal scale for nativeness. A species considered native at a broad regional level (such as the southeastern United States) is probably not locally adapted to all habitats within that range. Additionally, human activities have blurred historical distribution patterns, as species have been intentionally and unintentionally introduced beyond their original ranges. Another challenge is distinguishing native species from **naturalized nonnatives**—plants that, while introduced, have established self-sustaining populations without becoming invasive (that is, degrading native plant communities). The struggle over what constitutes a native plant has implications for conservation, restoration, and landscaping decisions, because prioritizing native species is often seen as a strategy for preserving ecosystem integrity and supporting wildlife. In some cases, however, nonnative species may also provide ecological benefits, suggesting that careful consideration, rather than an absolute ban, is warranted for noninvasive species.

Meadows in the Context of Ecological Restoration

If meadows are not naturally occurring in Florida, what is the benefit of introducing them? To understand this, it is crucial to distinguish between ecological restoration and ecological enhancement. **Ecological restoration** is the process of aiding the recovery of a natural community that has been degraded, damaged, or destroyed.

Restoration of natural communities is vital for the health and long-term survivability of Florida's flora and fauna. However, restoration is typically expensive, time-consuming, and not always feasible. True restoration also demands specific regional expertise, specialized equipment, and access to

Restoration from a pine plantation (*left*) to a pine savanna (*right*) is typical in Florida and represents a feasible restoration project that increases vegetation density and diversity. Photo by Gage LaPierre.

Ecological enhancement of a degraded pasture (*left*) to a wildflower (*right*) stand can be achieved with modest inputs when compared to the high inputs of restoring it to a pine savanna. Photo by Gage LaPierre.

appropriate plant material supplies for success, meaning it can be achieved more easily on certain sites than others.

Ecological enhancement or **rehabilitation**, by contrast, is an effort to modify or change a site to improve habitat for native flora, fauna, and people. The goal of enhancement is not to fully restore an area to its original indigenous plant community but instead to either partially restore or simply mimic a local natural community. Ecological enhancement is being more frequently employed as an alternative to restoration on sites where restoration is not physically or financially feasible. At times moderate improvements in wildlife habitat can be made with modest inputs. Therefore, the meadow concept described in this manual aligns more with ecological enhancement goals rather than true restoration.

Ecological enhancement acknowledges that many landscapes have been irreversibly altered and seeks to increase their ecological value within existing constraints rather than restoring them. Various approaches fall under the ecological enhancement umbrella: **Rewilding** emphasizes the return of natural processes and species to a degraded area; **rehabilitation** focuses on repairing ecosystem functions without necessarily restoring historic conditions; and **reclamation**, which is often used in postindustrial or heavily disturbed sites, seeks to stabilize the environment and support new plant and animal communities.[8]

A meadow installation at a residence in Micanopy, Florida. Strategic use of paths, plant textures, and layering creates a cohesive design. Photo by Matthew Wall Photography.

Meadow installation at Nemours Children's Hospital in Orlando, Florida (for a case study of this project, see chapter 5). Photo by Nancy Bissett, The Natives.

Why Meadows?

The installation of meadows is becoming increasingly popular throughout the state of Florida and the United States. Whether in municipal, commercial, rural, or urban residential spaces, meadows offer numerous benefits for people and the environment.[9] Chief among these is that meadows are an economically viable and sustainable alternative to intensively managed landscapes, such as lawns.[10] When designed, implemented, and managed properly, meadows can offer rich plant and animal diversity.[11] In turn, meadows can be used to help conserve biodiversity on a local scale and improve ecological resilience at a regional scale.[12] Aside from these benefits, meadows also tend to be more aesthetically pleasing and exciting than traditional landscapes.[13]

In Florida, meadows can be created across a variety of environments, soil conditions, and spatial scales. This includes fields, pastures, open or cleared woodlands, power line corridors, and roadsides, as well as residential yards and urban lots. In most cases, trees and woody plants have limited presence but are not entirely absent from these areas. Generally in Florida, human management through grazing animals, fire regimes, or mowing is required to restrict woody plant abundance in created meadows. The lack of woody species is important to limit shading and encourage a predominance of herbaceous flowering plants (forbs), as well as grasses and sedges (graminoids). With proper design, installation, and management, meadows can lead to floristically rich and diverse plant communities that benefit both people and wildlife.

Hence, meadows can be considered a form of ecological enhancement (rewilding, rehabilitation, or reclamation). In the context of meadows in Florida, ecological enhancement can take the form of converting underutilized or degraded lands into diverse, functional plant communities that support pollinators, improve water filtration and carbon sequestration, and contribute to the stability of local ecosystems. By incorporating these strategies, meadow creation represents a practical and impactful way to enhance ecological value in both urban and rural settings.

Economic Benefits

Meadows provide a range of economic advantages when compared to conventional landscaping with single-species plantings (monoculture) such as lawns or meticulously manicured flower beds. Foremost, meadows are usually cheaper to install and maintain on a per-acre basis.[14] Conventional landscapes typically require extensive preparatory work, including topsoil importation, fertilization, mulching, and the establishment of irrigation systems and weed barriers.[15] In contrast, most meadow installations only require site preparation for weed control before planting. The costs can vary widely, depending on the site and the goals associated with a project. Hiring a professional to install a meadow within an urban space typically costs between $950 and $4,000 per acre, whereas a conventional landscape installation runs between $4,000 and $12,000 per acre.[16]

Additionally, unlike traditional lawn monocultures, which often demand weekly care such as mowing, pesticide application, and regular reseeding, meadows thrive with infrequent mowing and minimal care.[17] This is because meadows are designed to mimic natural ecosystems, where diverse plant communities naturally compete, support each other, and adapt to local conditions, reducing the need for artificial interventions. Moreover, meadows eliminate the reliance on the chemical inputs such as fertilizers, fungicides, herbicides, and insecticides that are widely used in conventional ornamental landscapes. This not only reduces costs but also curbs the harmful environmental consequences associated with chemical runoff.[18] All in all, the average annual maintenance cost of conventional urban landscapes runs between $300 and $5,452 per acre, compared to $50 to $750 per acre for meadows.[19]

Water usage is an increasingly important economic factor where meadows excel. Lawns and nonnative ornamental landscaping in Florida often demand substantial irrigation, especially during the dry months from mid-April to mid-June, while meadows composed of native species are adapted to the local climate and require far less supplemental water.[20] Excessive water use for landscaping not only raises maintenance costs but can also lead to violations of water-overuse regulations, particularly in areas where water scarcity is a pressing issue. By transitioning to meadows, property owners and municipalities can significantly reduce their water consumption, aligning with sustainable water management practices while saving money.

The full scope of economic benefits associated with meadows is still emerging, but the advantages already recognized are compelling. Meadows require fewer fossil fuel inputs and chemical applications than traditional landscaping, which can mitigate broader socio-environmental and public health costs. For instance, the fertilizers and pesticides commonly applied to lawns contribute significantly to the degradation of water quality in Florida's waterways, including water contamination and nutrient overloading.[21] Nutrient overloading in waterways causes algae blooms and contributes to the growth of invasive plants, the removal of which costs Florida taxpayers billions of dollars annually.[22] Additionally, air pollution from the two-stroke engines frequently used in lawn maintenance equipment compounds environmental and health issues.[23] These types of machines emit harmful pollutants, including benzene and carbon monoxide, which are linked to cancer and respiratory illnesses. Unlike modern four-stroke engines, two-stroke engines lack a dedicated lubrication system, resulting in a significant portion of the fuel-oil mixture being expelled unburned. This design makes them notoriously fuel inefficient, with some studies indicating that they can emit as much pollution in a short period as a car driven hundreds of miles. This disproportionate pollution output relative to their size means that even brief use of these tools contributes substantially to localized air pollution and health risks. In stark contrast, meadows require minimal use of maintenance equipment because the diverse plant community naturally suppresses weeds and requires only infrequent mowing, often just once or twice per year.

The historically popular shift toward intensive landscape management practices, such as maintaining monoculture lawns, is also a major factor in

Overuse of fertilizers and pesticides has been linked to water contamination and nutrient overloading in Florida waterways. "Green Tick and Mosquito Spraying," photo by Praxis Eco Pest Control, 2016, Creative Commons license via Flickr, https://www.flickr.com/photos/praxisecopest/29605770121/.

Algae bloom in a lake as a result of nutrient overloading has caused fish die-off. P "Algae and Dead Fish in Dianchi Lake, China," photo by Greenpeace China, 2007, Creative Commons license via Flickr, https://www.flickr.com/photos/48722974@N07/4598769539/in/photolist.

the ongoing decline of pollinator populations across the southeastern United States.[24] Pollinators like our native bumblebees are vital to many agricultural systems because they facilitate fruit set and improve crop yields, so their decline jeopardizes agricultural production.[25] Alongside pollinators, many other insect and wildlife species in Florida are experiencing significant habitat loss due to conventional landscaping practices.[26] Specifically, the creation of monoculture greenspaces eliminates the diverse array of native plants that provides food and shelter for a variety of species. Furthermore, the heavy use of pesticides directly kills insects, which disrupts the food web, leading to a decline in biodiversity. The overall decline of insect populations poses long-term threats to both ecosystem stability and the economy, given the interdependence of agriculture, biodiversity, and natural resource health. Meadows, by contrast, offer essential habitat and forage for pollinators and other wildlife, promoting their recovery and supporting broader ecological resilience. They act as havens where biodiversity can thrive, underscoring their importance not only as a landscaping alternative but as a keystone of conservation efforts.

The Nationwide Turn to More Sustainable Greenspaces

The trend toward meadow creation is nationwide. In recent years the hunger for more sustainable and ecologically minded landscaping has shaped the way many landscape architects approach design in commercial, residential, and

Meadow landscapes of The High Line in New York City, emphasizing their aesthetic integration into urban environments: (*top left*) the elevated walkway of The High line filled with meadow plants; (*right*) a side view of The High Line meadow, showing the parking garage underneath; (*bottom left*) The High Line green space attracts many people in its urban environment. Photos by Isabella Guttuso Browne.

Meadow at Little Island in New York City. Photo by Isabella Guttuso Browne.

This meadow at Bellevue University is used for student research in several different disciplines. Photo by Benjamin Vogt.

The former lawn surrounding a parking lot at the University of Nebraska Medical Center campus in Omaha was converted into a meadow. Pictured in bloom are hoary verbena (*Verbena stricta*), prairie coneflowers (*Ratibida pinnata*), and wild bergamot (*Monarda fistulosa*). Photo by Benjamin Vogt.

public greenspace settings. Many of these landscape trends include a shift toward the use of more native plants and less turfgrass. **Meadowscaping** is often used to describe this new and increasingly popular approach to urban landscaping. The design philosophy was pioneered by landscape architects Wolfgang Oehme and James van Sweden, whose New American Garden movement replaced lawn-dominated landscapes with bold, layered plantings that celebrated texture, color, and seasonal change. Building on this legacy, garden designer Piet Oudolf and his New Perennial movement have brought herbaceous perennials and grasses into high-profile projects such as Lurie Garden in Chicago and The High Line and Little Island in New York City, bringing meadow-inspired design into the urban mainstream.

In the US South and Midwest, the idea of pocket prairies is starting to catch

Many native perennial grasses and forbs, such as Florida's native switchgrass (*Panicum virgatum*), have deep root systems capable of capturing carbon and stabilizing soils. Photo by Steve Renich, 2008, Creative Commons license via Wikimedia Commons, https://commons.wikimedia.org/wiki/File:Switchgrass_roots.jpg.

on in many urban and suburban areas. **Pocket prairies** are small, fragmented patches of prairie vegetation, often created in urban or suburban areas to mimic the native prairie ecosystems that once covered much larger areas. While the origin of the term is somewhat unclear, it appears to have gained popularity in conservation and ecological restoration circles in the early twenty-first century. The idea was likely inspired by the concept of pocket parks (small green spaces in urban areas) combined with an increasing interest in preserving or restoring prairie ecosystems. It reflects the idea that even small plots of land can contribute to biodiversity conservation, provide habitat for native species, and serve as educational tools for the public.

Ecological Benefits

Overall, meadows are far more ecologically friendly than most conventional landscapes and land uses.[27] Conventional landscaping frequently relies on a limited selection of horticultural plants. Oftentimes these plants are not native and do little to support native ecosystems. Meadows, in contrast, can support a high diversity of native plants that offer food and habitat for insects, pollinators, and wildlife. Meadows can also improve water purification, carbon sequestration, stormwater mitigation, and soil remediation.[28] Additionally, the diversity of plants in a meadow planting results in a range of growth periods, root architectures, functional groups, and other factors that greatly enhance the diversity of the soil biome, improving soil structure and health.[29]

Conceptually, meadows grant more opportunities to ecologically enhance sites that are not feasible to restore, such as pastures, construction debris sites, and abandoned agricultural fields. Thereby, meadows can play an im-

Meadows installed near swales, bioretention areas, or within rain gardens can help reduce flooding, improve water quality, and provide habitat for wildlife. Photo by Nancy Bissett, The Natives.

Meadows can provide valuable habitat for a variety of species. Here, a gopher tortoise has made its burrow within a meadow planting. Photo by Andrea England with My Florida Meadow Company.

portant role in providing buffers and linkages between existing conservation areas and in urban-to-rural corridors. Such corridors have become a focus in Florida due to rapid suburban sprawl and the need to preserve the integrity of remaining wildlife populations. When designed with intention, meadows can help in conservation efforts like these as well as the protection of endangered or threatened plant populations.

Increasing urbanization and land clearing throughout Florida and much of North America have caused dramatic declines in many populations of insects, birds, and other wildlife.[30] These trends are directly related to how Floridians have historically and are currently managing land. With an increasing population and looming effects of a changing climate, these trends are expected to accelerate. Installation of meadows in areas from small yards to large rural lots can help mitigate the effects of a changing climate and contribute significantly to the preservation of Florida's plants and fauna.

Pollinators Need Meadows Now More Than Ever

Native pollinators are on the decline in Florida due to habitat loss, overuse of pesticides, and over-management of our yards. Meadows are a chief means of supporting pollinators. Unlike traditional manicured lawns or mowed fields, meadows offer a varied and continuous supply of nectar and pollen throughout the growing season, catering to the specific needs of different pollinator species. The native wildflowers and grasses in meadows not only serve as a food source but also provide essential shelter and nesting sites for various insects, including solitary bees, moths, and butterflies. Solitary bees, for example, often create nests in bare patches of soil, which are readily available in a well-established meadow. Clumps of bunchgrasses and fallen leaves provide overwintering sites for butterflies and moths, while dense foliage offers refuge from predators and harsh weather conditions. Additionally, the varied heights and densities of plants within a meadow create microclimates, providing vital shade and humidity regulation for insects. Leaving standing dead stems and avoiding excessive fall cleanup further enhances these nesting and shelter opportunities. The introduction of meadows in residential settings contributes to the overall health of pollinator populations by min-

A butterfly visiting a milkweed plant native to Florida. Photo by Nick Freeman, Wacca Pilatka, LLC.

left A bee visiting a coneflower native to Florida. Photo by Nick Freeman, Wacca Pilatka, LLC.

middle A pollinator visiting Florida native wildflowers. Photo by Nick Freeman, Wacca Pilatka, LLC.

right A bee feeding on an ironweed plant (*Vernonia gigantea*) native to Florida. Photo by Nick Freeman, Wacca Pilatka, LLC.

imizing the use of pesticides and herbicides, fostering a more natural and sustainable environment. As homeowners embrace meadows, they actively participate in conserving biodiversity, creating a haven for pollinators, and promoting the essential ecological services these species provide, ultimately enhancing the overall health and resilience of local ecosystems.

Aesthetic Benefits

Studies have yielded evidence that humans are on average more at ease and drawn to open environments than to other natural settings.[31] This attraction to open environments is thought to account for the advent of the monoculture lawn aesthetic in Europe and the United States.[32] However, aesthetic values evolve, often in response to societal norms and pressures. Today, the growing urgency to tackle our biodiversity crisis is starting to influence our aesthetic and cultural values when it comes to how we manage land. Presenting people with the practical and ecological benefits of meadows can alter their perceptions and positively influence their aesthetic sense.[33]

Meadows offer a far richer and more diverse sensory experience than do lawns or most conventional landscapes.[34] A typical lawn provides only a single texture, a uniform green or brown color, and limited variation in height—short if mowed, tall if neglected. In contrast, meadows display a dynamic mosaic of colors, textures, and heights that shift with the seasons and evolve over time. The interplay of blooming wildflowers, waving grasses, and humming pollinators creates a living, ever-changing landscape that stimulates the senses.[35] Unlike traditional lawns, which remain mostly static and uniform, meadows are vibrant, embodying life and movement throughout the year. Their seasonal changes and capacity to support diverse wildlife and pollinator populations make them aesthetically rich and ecologically valuable.[36]

Even small meadows offer homeowners a unique canvas for personal aesthetic expression. Unlike the rigid uniformity of a lawn, a meadow allows for creative freedom in plant selection, layout, and maintenance. Homeowners can curate their own miniature ecosystems, choosing specific flowers for color palettes, textures for visual interest, and native plants to attract desired pollinators. The ever-changing nature of a meadow means that each season brings new opportunities for visual delight and creative design, transforming the landscape into a personalized work of living art.

A meadow in south-central Florida illustrates artistic and ecological beauty. Photo by Gage LaPierre.

Students of various ages engaged in educational activities in a meadow at the University of Florida Natural Area Teaching Laboratory (NATL). Photo by Gage LaPierre

Education and Teaching Benefits

Meadows may appear to be just aesthetic and ecological enhancements, but they hold immense potential as vibrant educational resources as well. These dynamic spaces provide a hands-on learning environment where students and educators can deepen their understanding of the natural world. Through activities like identifying native plants and observing the intricate interactions between flora and pollinators, meadows become living classrooms. They offer unique opportunities for conducting experiments, collecting data, and applying scientific methods in real-world settings, bridging the gap between textbook knowledge and practical application. Beyond science, meadows can inspire creativity and personal expression through nature journaling, sketching, poetry, and artwork. These activities not only nurture artistic talents but also cultivate an emotional connection to the natural world, fostering a sense of wonder and appreciation for its beauty, dramaticism, and complexity.

A child playing in the sand near a natural pond showcases the human connection to natural landscapes. Photo by Nancy Bissett, The Natives.

Students engaged in educational activities in a meadow at Bok Tower Gardens, near Lake Wales, Florida. Photo by Nancy Bissett, The Natives.

A well-designed meadow in Central Florida with a mix of native grasses and wildflowers, showcasing ecological landscaping. Meadow designed by Nancy Bissett. Photo by Nancy Bissett, The Natives.

As Florida continues to urbanize, its residents face an increasing disconnection from the natural world. Meadows, while not a replacement for wild spaces, serve as accessible sanctuaries that bring nature closer to home. They provide spaces for current and future generations of Floridians to learn about and connect with native flora and fauna, rekindling a sense of belonging to the landscapes that define the region. By fostering this connection, meadows help cultivate a sense of place and identity tied to Florida's unique ecosystems. Moreover, they can inspire greater sentiment and advocacy for the conservation of natural areas, both locally and across the state. In this way, meadows are more than just green spaces—they are gateways to education, inspiration, and stewardship that support a sustainable future for Florida's people and environment.

PREPARING TO CONSTRUCT A MEADOW

Planning and design are critical to successful installation and long-term development of a meadow. Purposeful site selection is not always possible, but a high presence of weedy nonnative plants or invasives is very difficult to remediate and will threaten the establishment of meadow species. The plan and design of a meadow partially depend on the goals and objectives being set for a site, because the goals of a project ultimately set expectations and define success. Since meadows are human-created landscapes, they are constructed following principles of landscape design to foster functional and aesthetic values.

Goals and Objectives

Goals for a meadow project can be broad but should encompass what you hope to see as an outcome. Once the overall goals are determined, then more specific measurable objectives can be crafted and planned accordingly. The goals of a meadow will differ depending on the context of the project. For example, land managers seeking to install a meadow in a large rural area might identify goals such as improving habitat for game species or reducing coverage of invasive species. For a residential homeowner, on the other hand, the goals for a meadow project may be to have a showy front yard that

opposite This beautiful meadow showcases the ecological and aesthetic diversity of meadows. Photo by Andrea England with My Florida Meadow Company.

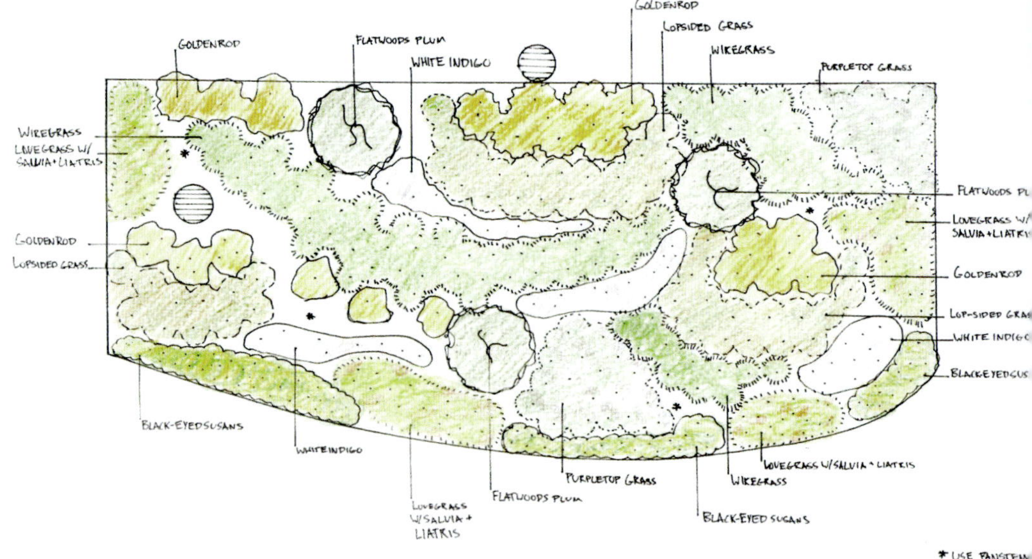

Meadow design for the University of Florida NATL project. Illustration by Blake West at Zone 9 Design.

supports pollinators and reduces inputs for turfgrass maintenance (that is, water, fertilizer, and pesticides).

A helpful way to determine goals and objectives for a site is to set desired target conditions. You might ask, How exactly should the site look once it is fully developed? For example, one target condition for an urban meadow could be to attain an ecologically beneficial and aesthetically pleasing site. Based on this goal, you can then zoom in on more specific objectives like achieving more than 85 percent coverage of a particular set of native herbaceous flowers and grass species with less than 5 percent coverage of undesirable nonnative species. From here a project manager can set firm benchmarks for when success has been achieved and be able to communicate what a site will look like when completed.

Meadow Design and Planning Process

Planning a meadow does not entail just the placement of plant material. The process of planning and designing a meadow involves a site-specific approach that acknowledges the diverse factors influencing each project site in Florida. This requires understanding all relevant variables, such as soil composition, sun exposure, drainage, climate, land-use history, and weed presence. Successful meadows also require consideration of the landscape context and neighboring features such as these:

- Will the meadow be part of a larger ecological enhancement effort near an intact natural community?
- Does the project neighbor an area filled with invasive plants?
- How do people nearby currently interact with the landscape where the meadow is planned?
- Is the project in a residential front or back yard that needs space for children to play?
- Is the setting a public park where visitors may want to walk through the meadow?

Asking specific questions like these is helpful when identifying the components of a meadow design that are key for reaching ecological, aesthetic, and functional goals.

Site Selection

The placement of a meadow is important, but often the location of a meadow installation is predetermined. In urban settings, for example, meadow sites are often selected based solely on space availability within the urban matrix, whereas in rural settings, they are often located in areas where full restoration is not feasible. Nonetheless, when site selection is possible, the preferred site qualities for a meadow are a flat or gently sloped, sunny location that features relatively low soil fertility and limited soil disturbance (that is, compaction, inversion, translocation). Disturbed soils with a pH greater than 7.5, or those with a high clay content at the ground surface are not ideal. Desirable sites may have trees present but not at a density where shady conditions will become dominant over time. In addition, no past or current presence of invasive plant species or dominance of certain nonnative species is preferred.

Unutilized spaces can be designed and converted into meadows. Photo by Gage LaPierre.

Urban landscapes can be transformed into meadows that provide both ecological benefits and visual beauty. Photo by Troy Springer.

left Pine trees in a meadow dominated by blazing star (*Liatris* spp.). Photo by Gage LaPierre.

top right Sabal palm trees in a North Florida meadow. Photo by Harriet Festing.

bottom right Live oaks and laurel oaks shading out understory vegetation. Photo by Gage LaPierre.

Notes on Trees in Meadows

Trees can greatly alter and influence meadow structure and composition. The impact on a meadow is greatly determined by the tree species. Live oak (*Quercus virginiana*), laurel oak (*Q. hemisphaerica*), and water oak (*Q. nigra*), for example, all produce high amounts of shade and leaf litter that can dampen desirable plant abundance or diversity. Far less impactful are pine-savanna oaks such as southern red oak (*Q. falcata*), turkey oak (*Q. laevis*), and sand post oak (*Q. margarettae*), as well as most native pines such as longleaf (*Pinus palustris*), slash (*P. elliottii*), and loblolly (*P. taeda*) pines. These species produce far less shade, drop litter that enables the site to burn more effectively, and do not smother it with debris over time. Small trees such as chinquapin (*Castanea pumila*), flatwoods plum (*Prunus umbellata*), red buckeye (*Aesculus pavia*), and persimmon (*Diospyros virginiana*) are great species to have within a meadow but should be planted along the margins, where debris will not interfere with mowing. Sparse use of cabbage palm (*Sabal palmetto*) also creates a favorable overstory for a meadow due to low litter and shade production.

Site Assessment

After a site has been chosen and goals for the project identified, the design process begins by performing a site inventory and assessment. In this stage, the context of the site is compiled, encompassing soil conditions, hydrologic conditions, slope, sun exposure, weed presence, and built features. The following is a list of criteria that can guide the design of a meadow prior to installation.

SUN EXPOSURE

Most plants associated with Florida meadows require full- to half-day sun. Thus, a space with full sun or little shade due to a tree canopy or built structures is ideal. When assessing a site, consider nearby tree lines that might grow over time and shade out the meadow. Also take note of the time of year when you are recording sun exposure, because lower light levels are common in many areas during the winter compared to the summer.

SLOPE AND HYDROLOGY

Analyze the topography of the location and the neighboring land. Note the direction of slopes and depressions. For larger sites, document the location and direction of waterways, wetlands, runoff points, hills, and mounds. Water requirements differ across plant species suitable for meadow planting, so this information is critical for selecting plants that will thrive in the conditions of the selected site. Mapping out locations where the terrain's slope changes is useful for guiding species placement across the site.

SOILS

Sites with low soil fertility and limited soil disturbance are preferred for meadow installations. Types of soil disturbance include **soil compaction** (the compression of soil particles, which reduces pore space and hinders root growth), **soil inversion** (the turning over of soil layers, which disrupts the natural soil profile), and **soil translocation** (the movement of soil from one location to another, which often alters its composition and structure).

Most Florida native upland soils are dominated by sand and have low fertility, and native plants associated with these soils are adapted to those conditions. However, many site locations, especially in residential or commercial settings in the urban matrix, contain disturbed or amended soils. In

Soil disturbance examples, with native soil cleared or filled over in new developments: (*left*) a lot being prepared for building, (*right*) soil being prepared for the property around a newly built house. Photo by Gage LaPierre.

these cases, the native soil has often either been translocated or filled over. A simple ribbon test can give a rough estimate of sand-silt-clay composition in the soil. A **ribbon test** involves moistening a handful of soil and squeezing it between your thumb and forefinger to form a ribbon. The length and texture of the ribbon indicate the soil's composition: A long, smooth ribbon suggests high clay content, while a short, gritty ribbon indicates more sand. High clay content in the topsoil may hinder seedling establishment, and site remediation may be necessary for meadow establishment. In less severe cases, an alternative may be to purposefully select certain plant species that are more tolerant of clayey soils.

As a rough means of identifying soil profiles, use a soil map of the general location from the US Department of Agriculture Natural Resources Conservation Service (NRCS) Web Soil Survey or the University of California's SoilWeb app.[1] You can then correlate soil series from the soil map to natural community types using table 1 in LaPierre et al., "Florida Soil Series and Natural Community Associations."[2] It is also useful to compare this information with the site's current soil profile to identify changes in composition over time.

HISTORICAL USE

It is often useful to know what the historical natural community would have been on a site, especially in rural settings. Locations with prior agricultural use, livestock grazing, or waste dumping are at risk for potential invasive plant seeds and degraded soil conditions. Soil type analysis and historical imagery can be used to determine previous natural communities. Historical

Historical aerial photograph of Gainesville, Florida, in 1937, now the site of NATL. Photo by the US Department of Agriculture. Courtesy of the University of Florida, George A. Smathers Libraries Digital Collections.

aerial images for the state are available through the University of Florida's Digital Collections.[3]

WEEDS AND EXISTING VEGETATION

Take careful note of the existing vegetation on the site. Document the plant species that are found across the target area and on neighboring locations. Highlight the dominant species present, as well as the occurrence of any invasive and nonnative species. If the selected site was recently cleared and is bare of plants, then look closely at the plant community composition of the nearest location with a similar soil type. This will help you estimate local weed pressure and what the site could potentially develop into. Ideally, monitor the selected site in both winter and summer seasons to accurately assess the presence of annual warm- and cool-season species. A seed bank test may also be helpful to identify potential weedy species (see procedure below). This helps predict what plants, especially weeds, may sprout in the future. Additionally, it is worth noting the presence of desirable native species, which can be marked for saving or relocation, where possible.

Seasonal vegetation composition differences on the same site in winter and summer: (*top*) wintertime landscape dominated with ryegrass and wild mustard, (*bottom*) summertime landscape dominated with nutsedge and bunchgrasses. Photo by Gage LaPierre.

Notes on Invasive, Nonnative, and Native Species

Native species are organisms that are indigenous to a given region or ecosystem because of natural distribution and evolution. **Nonnative** species are organisms that do not occur naturally in an area but have been introduced there by humans, either deliberately or accidentally. **Invasive** species are nonnative organisms that harm native ecosystems. The Florida Invasive Species Council (FISC) documents and ascribes plant species as invasive in the state of Florida but is not directly responsible for regulating invasive species (see FISC for details and an invasive plant list).[4] Other resources for determining whether a plant species is native or nonnative are the "Atlas of Florida Plants" online and "Studies in the Vascular Flora of the Southeastern United States."[5]

The past or current presence of any invasive species within a project area should be of high concern and, if possible, the site should be avoided alto-

left Cogongrass (*Imperata cylindrica*), an invasive plant species common in meadows, highlighting its dense growth habits. Photo by leaf0605, 2023. Creative Commons license via iNaturalist, https://www.inaturalist.org/observations/181570790.

middle Guinea grass (*Megathyrsus maximus*), an invasive plant species common in meadows. Photo by leaf0605, 2024. Creative Commons license via iNaturalist, https://www.inaturalist.org/observations/207708640.

right Torpedograss (*Panicum repens*), an invasive plant species common in meadows, highlighting its dense growth habits. Photo by Yann Kemper, 2023. Creative Commons license via iNaturalist, https://www.inaturalist.org/observations/194062790.

gether. This is primarily because the site's **seed bank** (the reservoir of dormant seeds within the soil) may contain large amounts of seeds or propagules from past populations of invasives. If it is not possible to avoid such a site, then see chapter 3 for site preparation techniques that can mitigate the threat of invasive species returning. However, some species like torpedograss (*Panicum repens*) and cogongrass (*Imperata cylindrica*) are incredibly hard to control once established because of their deep underground rhizomes; the difficulty of controlling these species without intensive methodologies cannot be overstated. In addition, the presence of certain nonnative species on a site may create challenges in seedling and plug establishment. Of particular concern are species such as bermudagrass (*Cynodon dactylon*), nutsedge (*Cyperus rotundus*), and hairy indigo (*Indigofera hirsute*).

left Smutgrass (*Sporobolus indicus*) is a problematic nonnative plant common in meadows. Photo by leaf0605, 2023. Creative Commons license via iNaturalist, https://www.inaturalist.org/observations/181570830.

middle Nutsedge (*Cyperus rotundus*) is a problematic nonnative plant common in meadows. Photo by Dave Ozric, 2024. Creative Commons license via iNaturalist, https://www.inaturalist.org/observations/236550441.

right Hairy indigo (*Indigofera hirsuta*) is a problematic nonnative plant common in meadows. Photo by Victor Heng, 2023. Creative Commons license via iNaturalist, https://www.inaturalist.org/observations/189497038.

Comparison of soil profiles between undisturbed (*left*) and disturbed (*right*) Apopka soil series in Gainesville, Florida. Photo by Gage LaPierre.

Understanding Soil Conditions

Determining the current condition of the soil is critical for planning a meadow project. This can be done by taking several samples of the soil across the site. Soil samples are best taken using a soil auger, corer, or punch tube, which keep the soil horizon intact. However, a shovel or post-hole digger will also work. For meadow construction a 10-inch depth sample is deep enough to provide helpful insights into the condition of the soil. Photographing soil samples can be helpful for recordkeeping. If samples are difficult to collect due to the hardness of the ground, that likely means the soil is compacted. Use a penetrometer or soil probe to measure and assess soil compaction. Also, on highly disturbed sites note the sand-silt-clay composition throughout the soil sample as the soil profile may have become mixed (see the "comparison of soil profiles" photograph). For a small fee, the University of Florida ANSERV Lab will test soil samples for fertility and pH.[6] This information may help inform site preparation and plant selection decisions (see chapter 3). Lastly, using the soil samples collected, try to edit the soil map you created earlier to reflect localized conditions. Doing so will allow for adjustments in plant and seeding placement.

left Soil seed bank testing to assess weed pressure. Photo by Gage LaPierre.

right Soil seed bank tests help to identify dormant weed species. Photo by Gage LaPierre.

A **soil seed bank test** is a simple way of measuring the composition of potential weeds waiting dormant on a project site. To do this first take soil samples evenly across the site, ideally a minimum of 10 per acre. Spread the collected soil samples onto individual trays per sample prepared with an even layer of substrate germination mix. Maintain the trays in a sunny greenhouse at a constant temperature in the 70°F–86°F range and water regularly. Once seedlings emerge, identify their species and take note of abundance.

Elements of Design in the Meadow Landscape

While it may seem appealing to let a meadow establish itself naturally, embracing the "randomness" often associated with the term **natural** is rarely the

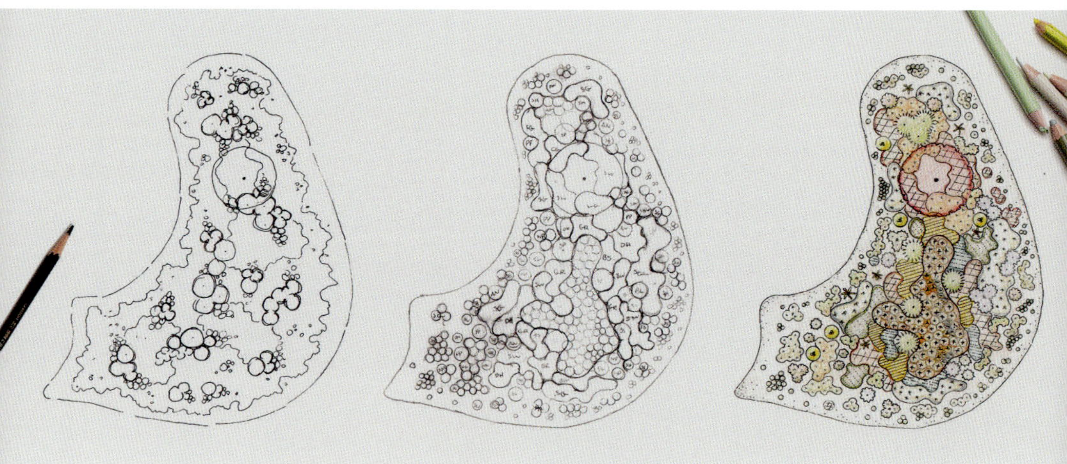

Process sketch for a meadow landscape, illustrating the layout of plants in a planting bed. Photo by Isabella Guttuso Browne.

ideal approach. The design of a meadow is a cornerstone of its long-term success, ensuring both ecological function and aesthetic value. However, there is no universal formula for a "good" meadow design, as every project is unique, shaped by the distinct characteristics of its site and its intended purpose. A successful meadow design hinges on a thoughtful evaluation of the project's goals, the needs of its users, and the insights gathered during site assessments. For instance, large rural meadows may favor a more hands-off or minimally structured design, while smaller urban and residential meadows often require careful planning to balance ecological function with aesthetic appeal and space constraints. Fortunately, several universal elements and principles of landscape design provide a foundation for designing all landscapes, including meadows. Four key elements—line, form, texture, and color—serve as the building blocks of a landscape. Guiding principles—proportion, rhythm, and function—direct how these elements are arranged to create a balanced and visually appealing landscape. The elements are the "what," and the principles are the "how." Understanding these ideas can help inspire and guide you as you design a meadow landscape. There is much that can be said about the elements and principles of landscape design, but the following portion of this chapter highlights key considerations, definitions, and examples particularly relevant to meadow projects.

Line

Understanding the role of lines is essential for creating a cohesive and functional meadow design. Lines serve as the framework that defines space, establishes patterns, and guides movement through the meadow. Depending on the project's stylistic goals, a design may incorporate geometric, structured lines for a formal appearance or embrace organic, flowing lines for a more natural and informal aesthetic. Common examples of lines in meadow design include bed lines, hardscape lines, path lines, and sod lines, all of which shape the meadow and influence how people interact with it. Soft, organically shaped edges and contiguous beds are often the most practical for maintenance, offering a seamless and manageable design. In more naturalized settings, meadows can also be allowed to transition gradually into adjacent plant communities, blending harmoniously with the surrounding environment and enhancing ecological connectivity.

No matter whether a planted meadow is geometric or organic in nature, lines are frequently dominated by the following aspects:

TRAILS AND PATHWAYS

The design of trails and pathways, especially those that bisect or run through the interior of a meadow, can affect user experience, meadow health, and maintenance costs. Ideally, interior or bisecting trails should not create disproportions in the overall layout of the meadow. The width of interior trails or pathways should also be appropriate for the user group. For example, small meadow installations or those with little foot traffic, like those in residential settings, may feature narrow foot trails of sod, stepping stones, mulch or another hard or soft trail material. Larger meadow installations may have a variety of trails, from narrow foot trails to wheelchair-accessible routes to significantly wider trails that can accommodate large groups and land managers moving through the site. Keep in mind that wide interior trails may increase maintenance costs and can potentially detract from meadow health (for example, invasive species introduction via mowing, increased edge effects). With that said, not all meadows need an interior trail or pathway. For example, trails and pathways around the exterior are typically ideal for small meadows, whereas in large meadows, an interior trail may offer maintenance and visitor interaction benefits. On the residential scale, pathways might not be necessary at all.

A gravel trail bisecting a meadow illustrates how trails can be an aesthetic component of meadow design. Photo by Nancy Bissett, The Natives.

BORDERS

Residential meadows benefit from edging or borders that separate the meadow from other spaces (such as walkways, play areas, gardens, turfgrass). Doing so makes it clear to visitors and users that the meadow is intentional and being maintained. Good-quality edging also helps keep out lawn grasses that would otherwise creep into the meadow over time. In residential areas, a metal or rock border may be appealing. For larger meadows hard physical edging may not be cost-effective. Instead, a soft border such as a mowed strip coupled with annual herbicide application or edging may prove more cost-effective. A mowed border is a simple, cost-effective

opposite A trail through a meadow at Bok Tower Gardens in Lake Wales, Florida, integrates organic, curvilinear lines that guide movement through the landscape and create visual contrast within the meadow. Photo by Isabella Guttuso Browne.

top Organic mowed edges are a classic border style for meadow landscapes. These clean lines help define order and signify that the landscape is well maintained, while allowing more natural or "wild" maturation of the meadow to occur within the bed. Photo by Matthew Wall Photography.

bottom left A meadow border using organic mowed edges. Photo by Heather Crane, Native Plant Horticulture Foundation.

bottom right A native Florida meadow is bordered by mowed turfgrass with a clean, organic edge. Note the deliberate design: from left to right, flowering grasses and wildflowers are framed by a mowed turfgrass edge that meets the pathway. Photo by Troy Springer.

top Pairing natural fencing with a mowed border defines space while gently guiding visitors to stay on paths, reinforcing design intent and a sense of order. Photo by Gage LaPierre.

bottom When using a hard border such as metal edging, stone, brick, or pavers, adding a buffer of mulch between the meadow plants and adjacent turf provides a visual transition and makes turfgrass maintenance easier. This softens the edge and helps prevent turfgrass encroachment. Photo by NaMa Native Landscapes.

option for larger meadows. Wood or plastic edging is another option but should be avoided in meadows where fire is intended to be used as a management tool. A meadow border may also be defined simply by existing turfgrass, another type of ground cover, or an adjacent pathway, which may be pervious (gravel, mulch, sand) or impervious (brick, pavers, concrete, or another form of hardscape).

HARDSCAPE INTERFACE

The transition from meadow to hardscape (such as concrete, brick, pavers, or asphalt) can be challenging to design. In most cases the simplest option is to use a mowed border that is edged periodically. This simple action will indicate to visitors that intentional care of the space is ongoing. Another option is to use a mulching material such as pine straw, shells, or bark to help soften the transition and keep plants from growing into the hardscape. In urban settings, meadows may also be contained by raised planters; although costly, this option is effective in containing meadow species and retaining soil.

left A raised Corten steel planter contains a bed of wildflowers. Photo by Isabella Guttuso Browne, Native Plant Horticulture Foundation.

right The edge of a stone path is softened by wildflowers. Photo by Evan Galbicka, designer at Emergent Gardens and Pulp Arts.

Hardscape-meadow transitions can use concrete or gravel to help manage plant spread and define spaces. Photo by Nancy Bissett, The Natives.

A variety of plant forms define the meadow landscape at Bok Tower Gardens near Lake Wales, Florida. Note the gradual height transition—from short grasses and yellow wildflowers in the foreground to taller grasses, shrubs, and trees in the background. Photo by Isabella Guttuso Browne, Native Plant Horticulture Foundation.

Form

Plant form—the shape and structure of plants in the landscape—plays a major role in how a meadow looks and feels. Grass species like bluestem (*Andropogon* spp.), switchgrass (*Panicum virgatum*), and the flowering spikes of goldenrod (*Solidago* spp.) or blazing star (*Liatris* spp.) have a tall, upright form that adds vertical emphasis to the meadow. Lower or mounding species such as dune sunflower (*Helianthus debilis*) and native blue porterweed (*Stachytarpheta jamaicensis*) fill gaps, blend planting areas, and soften edges. Plants with unusual shapes, like coralbean (*Erythrina herbacea*) or prickly-pear cactus (*Opuntia* spp.), provide contrast and can serve as natural focal points.

Plant form is not only about how a single plant is shaped or structured—but also about how plants are grouped together. Repeating species with similar forms throughout a meadow creates cohesion and flow, while selectively introducing contrasting forms adds visual variety and emphasis. The right mix depends on the goals for the meadow and is discussed further in the Principles of Design of a Meadow Landscape.

A variety of fine-textured plants create an airy front yard meadow that contrasts with a stone pathway. Photo by Evan Galbicka, Designer, Emergent Gardens and Pulp Arts.

Texture

Texture—the visual and tactile quality of plant surfaces—adds richness, contrast, and depth to a meadow. Like form, texture strongly influences how a landscape is perceived—from its visual interest to the way light and wind interact with plants, shaping the meadow's overall mood. Texture comes from both foliage and flowers: foliage texture is determined by leaf size, shape, thickness, and arrangement, while floral texture depends on the density and size of blooms, seed heads, and inflorescences.

Coarse-textured species have large leaves or bold structural forms that anchor the design and draw attention. Examples include fakahatchee grass (*Tripsacum dactyloides*), saw palmetto (*Serenoa repens*), and Adam's needle (*Yucca filamentosa*). These plants have strong visual weight and are effective as accents, borders, or focal features.

opposite Contrasting plant textures can add variety in native meadow landscapes. Photo by Heather Crane, Native Plant Horticulture Foundation

Three different forms—upright wildflowers, mounding grasses, and a spreading groundcover create height different in this front yard meadow. Photo by Isabella Guttuso Browne.

Medium-textured species act as transitions, helping different parts of the meadow blend together. Plants like black-eyed Susan (*Rudbeckia hirta*) and dotted horsemint (*Monarda punctata*) provide color and texture that connect coarse and fine layers, creating continuity across the planting.

Fine-textured species add softness, movement, and subtle detail. The thin leaves of fringed blue star (*Amsonia ciliata*), the delicate blooms and leaf structure of Florida pennyroyal (*Piloblephis rigida*), and the airy foliage and petite flowers of Leavenworth's tickseed (*Coreopsis leavenworthii*) lend a light, intricate quality to the meadow.

Texture also changes with the seasons. Some plants shift from coarse or medium textures to fine as they mature or bloom. For example, Elliott's lovegrass (*Eragrostis elliottii*) and purple lovegrass (*Eragrostis spectabilis*) display medium-textured foliage for much of the year, but in fall their airy, flowering plumes transform the meadow into a soft, fine-textured display.

Combining coarse, medium, and fine textures creates layers of visual complexity within the meadow. Each contributes differently coarse textures provide structure and emphasis, medium textures create transitions and unity, and fine textures lend softness and movement.

Unique form and course texture from the planter, fence, and plant material in this meadow make a bold statement. Photo by Evan Galbicka, designer, Emergent Gardens and Pulp Arts.

A warm late-spring display of color created by wildflowers and grasses in a range of hues. Photo by Evan Galbicka, Designer, Emergent Gardens and Pulp Arts

Fall in Florida brings bold displays of purple and yellow wildflowers, showcasing the state's signature autumn colors. Landscape by Troy Springer Environmental. Photo by Nancy Bissett, The Natives.

Muhly grass (*Muhlenbergia capillaris*) and purple lovegrass (*Eragrostis spectabilis*) feature bright green foliage for most of the year, but in fall they provide a striking seasonal display with pink and purple blooms

Color

Color is a vital element in meadow design, serving as both a visual anchor and an expressive feature that changes with the seasons. To harness the full potential of color, it's important to consider the seasonal dynamics of meadow species—including bloom times, duration of floral displays, and how colors evolve as plants mature, sport new growth, or go dormant. Thoughtful sequencing of species can ensure a continuous flow of color throughout the year.

Blooms are not the only source of color. Foliage contributes equally, providing interest long after flowering ends. The varying shades of green across meadow species can create subtle contrasts in tone and texture, especially during periods when flowers are less prominent. Some plants display foliage hues that extend beyond green, offering year-round visual appeal. For example, Elliott's lovegrass (*Eragrostis elliottii*) features a cool blue-green tone that adds contrast to surrounding plants, while narrowleaf silkgrass (*Pityopsis graminifolia*) produces silver-gray foliage and golden-yellow flowers that brighten fall meadows. Cardinal flower (*Lobelia cardinalis*) contrasts deep green to purplish foliage with a striking red bloom in more moist meadows, and white wild indigo (*Baptisia alba*) provides a bold white accent when in flower.

Color planning not only enhances aesthetic appeal but also supports ecological function—a diversity of floral colors attracts many kinds of pollinators and provides visual interest for people as well.

Principles of Design in the Meadow Landscape

With the foundational elements of meadow design in mind, the next step is to consider how the principles of design apply to meadows—the guiding ideas that bring line, form, texture, and color together. Three principles especially relevant to meadow landscapes are proportion, rhythm, and function. Applying these thoughtfully helps create meadows that are aesthetically pleasing, sustainable, and purposeful.

Proportion

The principle of proportion describes the relationship between the size, scale, and balance of elements within a meadow. Achieving the right proportion depends on the meadow's purpose, setting, and design intent. A large, open meadow may feature broad groupings of similar species to emphasize unity and a sense of expansiveness, while a smaller residential meadow might use variation in plant height and texture to create layering and visual interest within a compact space—and to meet goals such as providing privacy or enclosure.

Balance is another aspect of proportion. Symmetrical designs often convey formality and structure, while asymmetrical arrangements feel more natural and relaxed. Proportion can also be expressed through plant layering—placing taller species in the background, medium-height plants in the middle, and low or groundcover species in the foreground—to create depth, structure, and a sense of visual balance.

When proportion aligns with the meadow's goals—whether to feel spacious and open or enclosed and intimate—the design appears intentional, cohesive, and scaled to its setting.

Rhythm

Rhythm in a meadow design develops through the repetition and variation of line, form, texture, and color. It creates movement and visual flow,

top A meadow planted with large groups of singular, low-growing wildflower and grass species allows for an expansive view. Photo by Nancy Bissett, The Natives.

bottom Low grasses in the foreground create an open area, while taller species in the background create a backdrop for this urban courtyard space. Photo by Troy Springer.

guiding the viewer's eye through the space. Repetition can appear in many ways—through a recurring plant form, a repeating color tone, or a consistent textural pattern that unifies the design.

Gradual shifts in plant height, color, or texture can also build rhythm, adding a sense of progression across the meadow. For instance, transitions from fine to coarse textures or from warm to cool colors can suggest flow

A meadow landscape where the elements and principles of design come together to create a functional, resilient ecosystem that is both ecologically rich and visually appealing. Photo by Nancy Bissett, The Natives.

A mass planting of grasses is intentionally broken up with goldenrod. Photo by Troy Springer.

A naturalized meadow at Archbold Biological Station's Frances Archbold Hufty Learning Center offers natural beauty with a variety of colors, textures, rhythms, and other elements of design while using regionally native plants that thrive in the natural Sandhill ecosystem of Highlands, Florida, while offering visitors the opportunity to explore the landscape. Photo by Reed Bowman.

and direction without the need for formal structure. To keep the meadow visually engaging, rhythm can be intentionally disrupted with plants that stand out or contrast with their surroundings—a shrub in a sea of grasses or wildflowers, a uniquely shaped species, or a specimen tree that serves as a focal point or visual pause within the composition.

Rhythm connects movement through a meadow into a cohesive whole, creating a landscape that feels natural, balanced, and sequenced.

Function

Function ensures that a meadow meets not only aesthetic goals but also practical and ecological ones. A functional meadow is designed with its users and surroundings in mind, allowing for easy access, navigation, and maintenance while supporting the site's environmental objectives. Selecting plants suited to the existing soil, hydrology, and climate promotes healthy growth and long-term sustainability. Function also includes planning for management practices—such as mowing, prescribed burning, or reseeding—that help maintain the meadow's health and resilience over time. It emphasizes supporting natural processes like pollinator activity and seasonal change while creating spaces for people to experience and enjoy the landscape. When proportion, rhythm, and function work together, the result is a meadow that balances beauty, practicality, and ecological value—a meadow landscape that continues to evolve and thrive over time.

Plant and Feature Placement

The placement of plants is a crucial aspect of meadow design. The first key step in placement is to ensure that plants are selected and installed in areas that match the abiotic (nonliving, chemical, and physical) conditions necessary for their success. These **abiotic conditions** include factors such as sunlight exposure, soil type (including pH, texture, and nutrient content), moisture levels, and temperature. For example, some plant species need full sun, whereas others tolerate partial sun. In addition, certain species perform best in very dry soils, while others prefer wetter conditions. Understanding the results of a site evaluation is critical when confirming plant placements.

The next step is to consider the elements and principles of design discussed earlier—your preferences in form, color, texture, and line, and the overall goals or principles that you want the meadow to express. With these considerations in mind, you can design direct-seeding and planting zones within the meadow. Using containerized plants allows you to plan out the locations of plants individually or as mass plantings. When selecting where individual plants should be placed, consider the vertical flowering height of each species and the relative density of surrounding vegetation. For example, coupling shorter height grasses such as native species of lovegrass, tridens, or bristlegrass (*Eragrostis* spp., *Tridens* spp., or *Setaria* spp.) with taller flowering forbs such as native species of ironweeds, goldenrods, or blazing stars (*Veronia* spp., *Solidago* spp., or *Liatris* spp.) may reduce weed pressure and provide structural support for tall stems.

Plant Massing Technique

Mass planting, also referred to as planting in drifts or en masse, is a fundamental landscape design technique aimed at creating impact, drama, or harmony by filling a space with groups of the same plant. Mass planting is essentially the clustering of plants together, creating a puzzle of multiple different plant groupings. This technique is ideal for use in meadows, as large masses of plants can accentuate the color, texture, and other visual qualities of each plant. Over time, some meadows will naturalize, and plants may spread beyond the original mass boundaries. This helps soften the initial boundaries and planting patterns in the landscape. These plant massing areas can be filled with plugs, potted plants, or seed, depending on what is

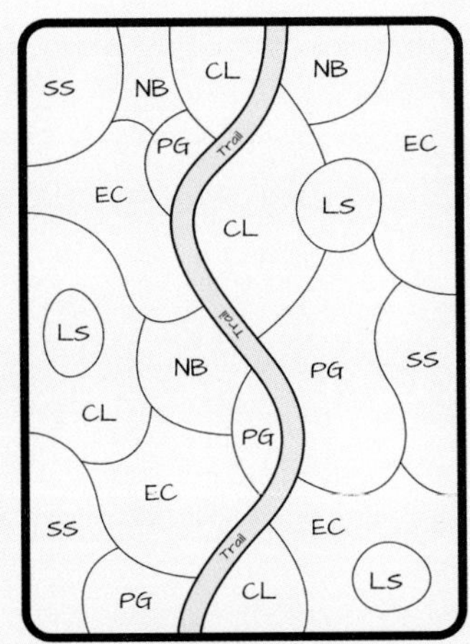

Example of a meadow design that uses the plant massing technique. Illustration by Isabella Guttuso Browne.

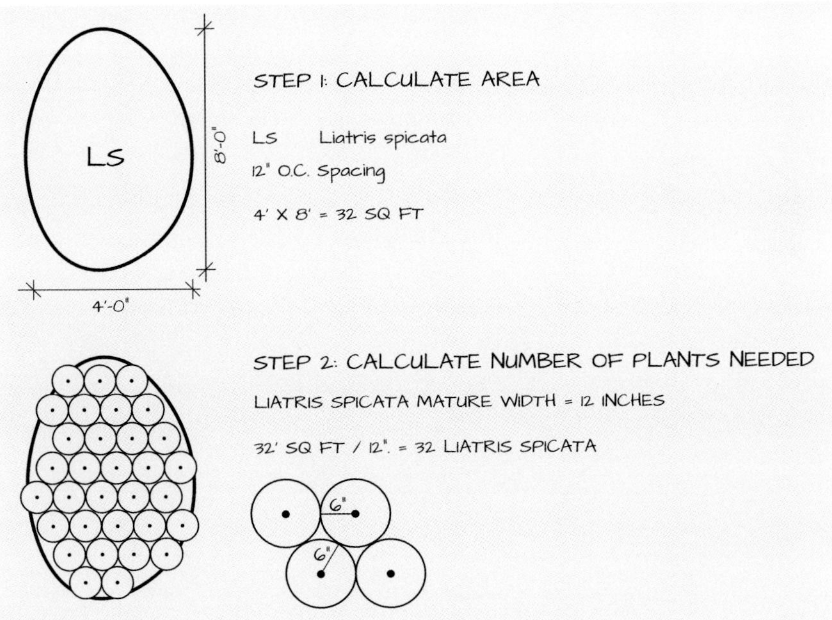

An example showing how to calculate the number of plants needed for a mass planting area based on each plant's mature width. Illustration by Isabella Guttuso Browne.

Plant layering using plants of different heights creates depth and structure in a meadow landscape. Illustration by Isabella Guttuso Browne.

readily available. If using plugs or potted plants, determine the number of plants needed by calculating the approximate area of each mass and dividing that by the mature width of the plant species. For smaller meadows, you can identify the location of each plant in the landscape.

Plant Layering Technique

Plant layering is another fundamental landscape design technique in which plants are categorized by height and size to create a layered effect on the landscape. Traditionally, layers consist of ground cover, low- to midsized shrubs, and specimen plants that are added as accents. Think in layers when considering different forms, colors, textures, and sizes of grasses, wildflowers, and other species suitable for meadow designs so as to accentuate their unique characteristics by the way they are placed.

Site Features

Depending on the goals for the site, other features can be incorporated into the landscape to enhance functionality. For example, furniture like benches facing the meadow can create places of rest and reflection. In publicly accessible landscapes, interpretive displays can add educational opportunities to

top left Cues to care are important elements that help bring visual order to a natural garden and signal intentional maintenance. Photo by Benjamin Vogt.

bottom left Standing deadwood (snags) serve as ecological and visual features. Photo by Harriet Festing.

right A shady tree in the meadow landscape can be used to provide a place of rest for visitors: a bench to sit on, a picnic table, or as in this photograph, a place to nap in a hammock. Photo by Heather Crane, Native Plant Horticulture Foundation.

left Educational signage in the meadow landscape can offer context and information for visitors about plant species selection and maintenance. This sign at the High Line in New York City explains the maintenance schedule for the meadow. Photo by Isabella Guttuso Browne.

right A learning garden in St. Augustine features tall educational plaques for plant identification. Photo by Nick Freeman, Wacca Pilatka.

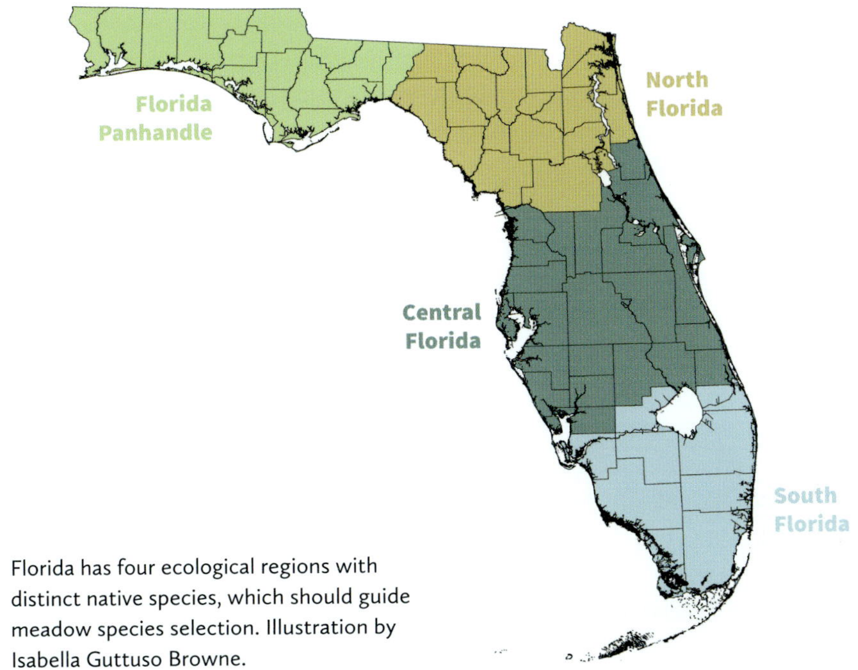

Florida has four ecological regions with distinct native species, which should guide meadow species selection. Illustration by Isabella Guttuso Browne.

the meadow. Typically, one might see plant labels identifying plants or signs offering information about the process of meadow management. Fixed outdoor art and rocks are another optional addition to enhance aesthetic appeal. These site features, which are easily recognizable as designed elements, are also known as **cues to care** that signal the landscape is being intentionally maintained.

Species Selection

The plant assortment in natural savannas and grasslands may appear random to the untrained eye; however, this is not entirely true. The composition of wildflowers and grasses in a natural area is frequently governed by abiotic factors such as soil conditions, hydrology, or disturbances like fires and herbivory (plant consumption) by wildlife. Additionally, in wild areas abiotic factors themselves often vary across a landscape, which helps to produce diverse plant communities over time.

Soil greatly determines which species will succeed or fail in a location. It is undoubtedly the most important abiotic factor to study when designing and selecting species for a meadow. On a relatively undisturbed site, the type of soil alone can be used as a filter in species selection. For example, dry-loving species such as lanceleaf tickseed (*Coreopsis lanceloata*), lopsided Indiangrass (*Sorghastrum secundum*), and sweet goldenrod (*Solidago odora*) are likely to thrive on deep sandy soils like those around Candler in Marion County, Florida. On poorly drained soils like those around Basinger in Okeechobee County, Florida, wet-loving species such as Leavenworth's tickseed (*Coreopsis leavenworthii*), sugarcane plumegrass (*Saccharum giganteum*), and Frank's sedge (*Carex frankii*) may fare better. Highly disturbed sites are tricky and require investigation into soil composition (pH, clay-silt-sand layers, organic matter). The soil profile of disturbed soils often varies substantially across a given site, especially for sites in former construction or agricultural areas. In the worst cases, the soil may have to be remediated or removed prior to planting.

Another crucial determinant for species selection is the geographical location of the project site within the state. In general, the state can be broken down into four ecological regions: Panhandle, North, Central, and South.[7] Due to differences in abiotic factors (such as freezing days, soil, and solar radiation) each region differs in the natural occurrence of native species and the growing range of nonnatives. Ideally, when designing a meadow, favor species that are native to the location's ecological region. These species are more likely to establish, persist, and support local pollinators and wildlife. However, species selection may be complicated by limited availability of regionally specific seed. Start by screening species according to site soil conditions, then according to ecological regional nativity, and finally filter species selection to meet ecological and aesthetic goals.

Meadows Need Native Graminoids

In most cases land managers and homeowners seeking to install a meadow are replacing a nonnative grass lawn monoculture. In this context, many people tend to place a lesser value on **graminoids** (grasses and sedges) in comparison to showy flowering forbs. However, a healthy meadow should contain a moderate coverage of graminoids (more than 40 percent), specifically native **cespitose** (bunch-forming) grasses. Native bunchgrasses and native sedges

Meadows grasses provide structure for forbs (flowering plants) and reduce weed influx. Photo by Gage LaPierre.

In this meadow grasses provide structural support for flowering plants, reduce bare ground, and promote wildlife habitat. Photo by Gage LaPierre.

are desirable because they provide structural support for many wildflowers as well as deterring weed encroachment by reducing bare ground. Native graminoids also provide aesthetic appeal during the winter months when showy forb species are dormant. Additionally, during the winter season, wildlife, namely birds, rely on the seeds and habitat cover that graminoids provide. Finally, graminoids also provide the necessary fuel if burning is to be used as a management tool. However, when installing a meadow, grasses combined should never make up more than roughly half of the seed mixture and ideally would be installed as plugs to ensure even establishment across a site.[8] In small-sized meadows, be judicious in use of taller bunchgrasses such as Fakahatchee grass (*Tripsacum dactyloides*) or switchgrass (*Panicum virgatum*) because they can obstruct views and crowd out other species. Shorter bunchgrasses such as wiregrass (*Aristida beyrichiana*) or Elliott's lovegrass (*Eragrostis elliottii*) would be better choices because they are short enough to avoid competing with most wildflower species.

Benefits of Native Legumes

Native legume species like partridge pea (*Chamaecrista fasciculata*), rabbitbells (*Crotalaria rotundifolia*), wild white indigo (*Baptisia alba*), hairy bush clover (*Lespedeza hirta*), beggarweed (*Desmodium floridanum*), and coral bean (*Erythrina herbacea*) are an important component of native ecosystems. All legume species rely on insect pollination, with most depending on bees for successful pollination. Following pollination, many legume species produce an abundance of seeds, which typically have a hard, water-impermeable coat that allows them to remain viable in the soil for years until conditions are suitable for germination. The seeds, flowers, and foliage of native legumes are an important food source for insects, birds, and mammals. Native legumes also enhance soil fertility through their ability to fix atmospheric nitrogen, improving the nutrient content of the soil. This nitrogen-fixing capability benefits not only the legumes themselves but also neighboring plants, contributing to the overall health of the ecosystem. Additionally, native legumes offer shelter and nesting sites for various species. The dense foliage of some legume species provides cover for small mammals and birds, while their flowers attract pollinators, supporting biodiversity.

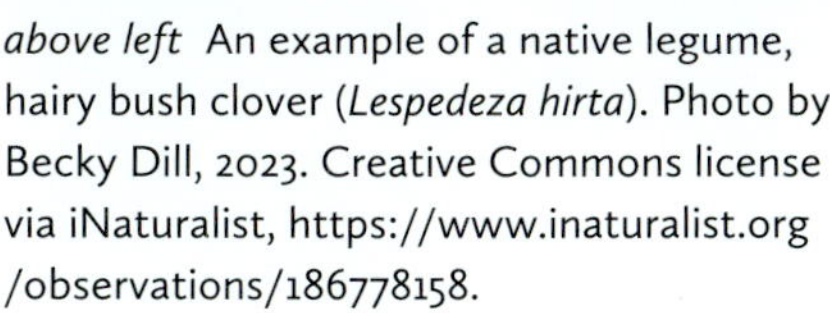

above left An example of a native legume, hairy bush clover (*Lespedeza hirta*). Photo by Becky Dill, 2023. Creative Commons license via iNaturalist, https://www.inaturalist.org/observations/186778158.

top right An example of a native legume, wild white indigo (*Baptisia alba*). Photo by Kirill Levchenko, 2024. Creative Commons license via iNaturalist, https://www.inaturalist.org/observations/225853021.

middle right An example of a native legume, partridge pea (*Chamaecrista fasciculata*). Photo by bogcheetos, 2023. Creative Commons license via iNaturalist, https://www.inaturalist.org/observations/172845848.

bottom right An example of a native legume, coral bean (*Erythrina herbacea*). Photo by Brady Reed, 2023. Creative Commons license via iNaturalist, https://www.inaturalist.org/observations/153391166.

top An example of a native cultivar (nativar), "Hello Yellow" (*Asclepias tuberosa*). Photo by Qwertzy2, 2005. Creative Commons license via Wikimedia Commons, https://commons.wikimedia.org/wiki/File:Asclepias_tuberosa.jpg.

bottom left An example of a native cultivar, "Uptick" (*Coreopsis* hybrid). Photo by PumpkinSky, 2017. Creative Commons license via Wikimedia Commons, https://commons.wikimedia.org/wiki/File:Coreopsis_tinctoria_cultivar_Uptick_Cream_and_Red_3.jpg.

bottom right An example of a native cultivar, "Indian Summer" (*Rudbeckia hirta*). Photo by Qwertzy2, 2005. Creative Commons license via Wikimedia Commons, https://commons.wikimedia.org/wiki/File:Rudbeckia_hirta_Indian_Summer.JPG.

Native Cultivars

Both cultivated varieties of native species, commonly known as **nativars**, and native-nonnative hybrids are becoming more common at nurseries and big-box stores. Nativars and hybrids are typically bred for specific human-desired traits such as flower size, duration, color, and growth height. Their use in meadows may help achieve aesthetic goals, with a few cautions. First, many lack long-term resilience against management techniques such as mowing and fire disturbances. Second, in many cases nativars and hybrids are ecologically benign. In some instances, however, their cultivated traits interfere with ecological traits that make the plants desirable for pollinators or wildlife. In rare cases, they may interbreed with wild populations as well, which is thought to diminish the local genetics of smaller populations over time.

In most cases the ecological goals of a meadow installation are to support native wildlife and pollinators through the creation of suitable habitat. Native plants in general support native wildlife more broadly and effectively than do nonnative plants. Native plants are also typically better suited to local climate, disease, and pest conditions. Additionally, nonnative species that do perform well in Florida—like zoysiagrass (*Zoysia* spp.), rhizoma perennial peanut (*Arachis glabrata*), and Asiatic jasmine (*Trachelospermum asiaticum*)—may eventually crowd out native species in meadows, limiting biodiversity. Thus, when constructing a meadow, the establishment of a diverse plant community dominated by regionally native species is the benchmark for achieving ecological goals.

On the other hand, not all native plant species are desirable in meadows, nor are all nonnatives ecologically unfavorable. Ruderal, or pioneer, native plants such as ragweed (*Ambrosia artemisiifolia*), dogfennel (*Eupatorium capillifolium*), and horseweed (*Conyza canadensis*) are extremely abundant throughout much of Florida, especially in disturbed areas. These species may offer structural habitat for some insects and wildlife, but they offer little else. In meadows they can also quickly dominate and overcrowd space that other, more beneficial, native species could occupy. Meanwhile, some nonnative species like blanket flower (*Gaillardia pulchella*), goldenmane tickseed (*Coreopsis basalis*), and annual phlox (*Phlox drummondii*) are less likely to overcrowd meadows. These species also appear to provide habitat and nectar for generalist pollinator species, although further research is needed to compare their benefits to those of native species.

An example of a ruderal plant, ragweed (*Ambrosia artemisiifolia*). Photo by Étienne Lacroix-Carignan, 2024. Creative Commons license via iNaturalist, https://www.inaturalist.org/observations/242525910.

An example of a ruderal plant, horseweed (*Conyza canadensis*). Photo by Étienne Lacroix-Carignan, 2024. Creative Commons license via iNaturalist, https://www.inaturalist.org/observations/255698798.

left An example of a fast-growing ruderal plant, dogfennel (*Eupatorium capillifolium*). Photo by Nonbinary-Naturalist, 2022. Creative Commons license via iNaturalist, https://www.inaturalist.org/observations/115018282.

right An example of a ruderal plant, Spanish needles (*Bidens alba*). Photo by Nonbinary-Naturalist, 2024. Creative Commons license via iNaturalist, https://www.inaturalist.org/observations/247332949.

Weedy Ruderal Plants

In ecology, **ruderal plants** are defined by their ability to quickly colonize and dominate recently disturbed areas. These are often fast-growing, opportunistic plants that thrive in areas where the soil has been disrupted by construction, clearing, or other disturbances. Think of them as the "first responders" of the plant world that quickly fill in bare patches. Examples of ruderal species commonly found in Florida include Spanish needles (*Bidens alba*), ragweed (*Ambrosia artemisiifolia*), cudweed (*Gnaphalium* spp.), thorny amaranth (*Amaranthus spinosus*), and many species of crabgrass (*Digitaria*). Ruderal species encompass both natives and nonnatives. These plants are well adapted to rapidly exploit disturbed soils and can quickly become dominant if not managed. Over time ruderals tend to be filtered out of plant communities in favor of longer-lived species due to competition or stress inducing-factors like fire, herbivory, or hydrology. Almost all Florida meadows will contain some ruderal species, especially early in their construction. The overdominance of ruderal plants, however, may limit the diversity of a site and hurt aesthetics. Therefore, careful management of ruderal species is necessary to

Bloom periods of wildflower species in a meadow landscape, emphasizing seasonal diversity for pollinator support. Illustration by Isabella Guttuso Browne.

promote the establishment of a diverse and balanced meadow community. See appendix A for a list of native ruderal species.

Bloom Periods

Another important factor to consider when selecting species for a meadow is the bloom periods across different species. Healthy meadows that are beneficial for pollinators should include as many different species and bloom times as possible. There are a wide variety of pollinators in Florida landscapes, including bees, wasps, butterflies, hummingbirds, moths, beetles, and even flies as well as bats. Each of these pollinators has evolved associations to particular sets of forbs, and even graminoids in some cases. Bear in mind that most common wildflowers in Florida tend to bloom in the spring and fall. Certain species, including Spanish needles (*Bidens alba*), blanket flower (*Gaillardia pulchella*), and scarlet sage (*Salvia coccinea*), can flower almost all year depending on the location. However, these species do not universally support all pollinators, and some aggressive species like Spanish needles (*Bidens alba*) can overcrowd other species in a meadow. Thus, if the goal of a meadow is to support pollinators, a diversity of wildflower species and bloom times is an especially important consideration.

Commonly Used Nonnative Wildflowers

The Department of Transportation has extensively seeded annual phlox (*Phlox drummondii*) and goldenmane tickseed (*Coreopsis basalis*) throughout North and Central Florida to create roadside wildflower stands. Native to eastern Texas, these species flower during the early spring months then reseed and **senesce** (wither and decline) when hotter temperatures kick in. Blanket flower (*Gaillardia pulchella*) was considered native to Florida until 2021.[9] Although it has been identified as nonnative, blanket flower does not appear to be harmful to native plant communities. On the contrary, it is very attractive to generalist pollinators and establishes exceptionally well along drier upland sites. Each of these species establishes well when direct seeded and is capable of coexisting with turfgrass. Moreover, they are readily available and affordable in the marketplace for bulk seed purchases.

When it comes to meeting aesthetic goals, height, texture, and color are the driving factors for species selection. Aesthetic concerns are typically greater in highly managed, people-dense urban and residential areas as opposed to rural settings. Ultimately, no single set of aesthetic qualities will please everyone. With that said, there is one single factor that will displease most people—neglect. Proper design and management should demonstrate thought and care of the space.[10] The following are some general points of guidance for species selection to meet aesthetic goals and mitigate the appearance of neglect:

- Do not select or allow common ruderal forbs such as ragweed, dogfennel, or horseweed to dominate. These species can overcrowd the site and may cause it to appear unkept.
- Consider how species will grow into the area over time. Small spaces are more subject to overcrowding by tall and large species. Thus, the smaller the space, the fewer tall and large species it should contain.
- Limit the use of thorny species such as blackberry (*Rubus* spp.), four-valve mimosa (*Mimosa quadrivalvis*), purple thistle (*Cirsium horridulum*),

Common nonnative wildflowers such as Annual phlox (*Phlox drummondii*) add vibrancy and adaptability. Photo courtesy of the Florida Department of Transportation.

An example of a common nonnative wildflower, goldenmane tickseed (*Coreopsis basalis*). Photo courtesy of the Florida Department of Transportation.

A meadow full of the nonnative blanket flower (*Gaillardia pulchella*) adds a visual element to a highway. Photo courtesy of the Florida Department of Transportation.

and most species of greenbriar (*Smilax* spp.), as well as species that sting like stinging nettle (*Cnidoscolus urens*), and heartleaf nettle (*Urtica chamaedryoides*). High coverage of these species may cause the meadow to appear unpleasant to visitors.

- Greatly limit aggressive vines that can overtake the site. These include muscadine (*Vitis rotundifolia*), greenbriar (*Smilax* spp.), peppervine (*Nekemias arborea*), and Virginia creeper (*Parthenocissus quinquefolia*). Again, excessive coverage of such species may produce an unmanaged appearance.
- When massing plants in smaller meadows (less than 800 sq. ft.) short or fine-textured plants such as lanceleaf tickseed (*Coreopsis lanceolata*), lovegrass (*Eragrostis* spp.), and helmet skullcap (*Scutellaria integrifolia*) may prove more appealing than taller or coarse-textured species such as ironweed (*Vernonia gigantea*), partridge pea (*Chamaecrista fasciculata*), or switchgrass (*Panicum virgatum*).
- A gradual increase in vegetation height from the meadow edges to the center or to distant edges may prove more attractive than placing tall species along the immediate margins, especially when adjacent to hardscapes or lawns. Mulching the edges of a meadow with pine straw, shells, or rocks is a good way to ease the transition from edges.
- To ensure color throughout the year, be sure to select a semi-even arrangement of spring-, summer-, and fall-flowering species.
- Consider creating clusters of flowering forbs for showy curbside appeal, which will simultaneously attract certain pollinators that favor such areas.

INSTALLING A MEADOW

Installing a meadow is a challenging process that combines science, art, and experience. For every installation project, carefully consider the steps needed to address site preparation, material installation, and post-installation actions. Of these phases, site preparation is by far the most critical to the success of meadow installation. Too often it is the one step most inadequately addressed. In many cases, site preparation will cover multiple seasons and thus should be a focus early in the planning phase. This will help mitigate shortcomings and leave open more options.

The top priority in every meadow project is to achieve the desired plant community composition. Weeds and noxious plants can threaten a project by outcompeting desired species early on. Invasive species that spread into one area of the meadow can also spell disaster by rapidly expanding to overtake the site. One goal of site preparation is to reduce or eliminate these species from the site prior to installation. Another goal of site preparation is to foster positive physical conditions for plant and seedling establishment. Poor site preparation is the primary reason meadow projects struggle or fail.[1]

Site Preparation to Enhance or Modify Physical Conditions

Some projects will require modification to the site's abiotic conditions for successful plant establishment. Recent construction sites and urban lots, for instance, may have extensive soil disturbance that requires remediation. This disturbance may involve soil compaction or the loss of native soil profile. In most cases remediating such sites will involve bringing in soil or fill material from off-site that will suit the desired plant materials. Sites that feature extreme soil compaction are difficult to remedy. One method that

opposite Blazing star (*Liatris*) in a meadow. Photo by Lilly B. Anderson-Messec.

Use of a cultipacker to prepare ground for seeding. Photo by Nancy Bissett, The Natives.

A residential site following herbicide treatment. Photo by Nancy Bissett, The Natives.

A tractor-drawn seed drill can be used to sow seeds on larger sites. Spiked rollers are at the bottom back of the vehicle and a container for the seeds of the same width sits on top of the frame attached to the rollers. Photo by Nancy Bissett, The Natives.

Tractor-drawn pan removing soil from a project site. Photo by Nancy Bissett, The Natives.

Flatwoods soil being spread across a project site. Photo by Nancy Bissett, The Natives.

can help is known as **subsoiling** or **soil ripping.** This involves the use of a tractor subsoiler implement that loosens compacted areas at depths below the soil surface.

In cases where it is necessary to import soil due to factors like extremely compacted or clayey soil, sterile soil is preferable because it reduces the risk of introducing new weed seeds. Mined sand with low clay content mixed with Florida peat or fine compost is ideal. In general avoid reclaimed soil or fill dirt due to the potential for contamination with weed seeds. An exception is the rare case where topsoil from an intact, natural area is being destroyed and can be salvaged and transplanted, a process known as **soil translocation**.

Clear plastic sheeting is used on-site for solarization.
Photo by Dr. Tyler Carney.

Site Preparation for Weed Control

A variety of site preparation methods exist to help eliminate weeds. There are pros and cons to each method, but they all can be used across most of Florida. Selecting the best method for a given project essentially comes down to resources available, site characteristics, and preference. Certain methods such as solarization and sod removal, for example, are more practical on small sites such as residential lots, whereas on larger sites herbicide, moldboard plowing, or repetitive cultivation may be more cost-effective.

Solarization

Solarization involves using clear plastic to heat the soil to temperatures that kill plants and weed seeds within the first 4 inches of topsoil. For this technique to work effectively, several things need to happen:

1. The plastic needs to be adequately sealed along the perimeter of the site to keep airflow out. This can be achieved by trenching the plastic into the ground or sealing the edges with dirt.

Plastic laid on-site for solarization helps to eliminate weeds and seeds.
Photo by Dr. Tyler Carney.

2. The project site should be moist prior to installing the plastic. If the site is too dry, seeds will be less affected by the high temperatures.
3. Solarization should be performed during late spring or early summer months to ensure the highest temperatures are reached.
4. A minimum period of six weeks of coverage is required for this process to be effective at reducing weed seeds in the soil.

With that said, solarization will not eliminate all weed seeds from the soil. Several species, including yellow nutsedge (*Cyperus esculentus*), are somewhat resistant to solarization. Nevertheless, solarization will usually kill off enough weeds to adequately prepare a project site.

STEPS FOR SOLARIZATION

1. **Spring:** Prior to installing plastic, you will need to cultivate or mow the site. **Cultivation**, in this context, means to clear the site of existing vegetation and till the soil. Cultivating the site with a rototiller is ideal because doing so largely eliminates perennial weeds. The site also needs

to be relatively flat, smooth, and clear of debris that can poke holes in the plastic. At this time, decide how the plastic will be secured along the edges to restrict airflow. On larger sites (greater than 1,000 sq. ft.) a trencher or tractor implement can make quick work of sealing the sides. On smaller sites the plastic can be dug into the ground using hand tools.
2. **Early summer:** At a time when the soil is moist, not wet or dry, spread the plastic over the site and secure the edges.
3. **Summer:** After installing the plastic, be sure to check for holes periodically. Seal any holes with tape to maintain high soil temperatures. Be sure to mow around the site and keep weeds away from the edges of the plastic to ensure it remains intact and to prevent weed seeds from setting.
4. **Fall:** Remove the plastic in the fall and follow promptly with seed and plant installation.

TIPS FOR SUCCESSFUL SOLARIZATION

- Clear plastic is preferable to black plastic because it allows sunlight to pass through, heating the soil to much higher temperatures.
- Walk-behind trenchers can be rented and are easy and clean to use. Trenches should be dug to a depth of at least 4 inches.
- Plastic should be at least 0.15 inches in thickness. Thinner plastic will puncture too easily and will generally degrade after one season.
- The plastic used in high tunnels is an excellent option for solarization of larger sites. It can often be found as used material in acceptable condition for solarization.
- To spread heavy rolls of plastic on a large site more easily, insert a pipe in the center of the roll and attach the ends to a tractor or all-terrain vehicle (ATV) to pull out the plastic.
- Do not place plastic within the dripline of trees, as this can harm the roots and potentially kill the tree.

Repeated Shallow Cultivation

The basic idea here is to repeatedly cultivate a given area to a shallow depth that allows germination of seedlings within the surface of the topsoil. Each round of cultivation kills new weed seedlings and brings new seeds to the soil surface. Repeating this process over an entire season, or preferably over

Rototiller cultivator for loose soil. Photo by Gage LaPierre.

Sweep cultivator for shallow weed removal. Photo by Magicman710, 2008. Creative Commons license via Wikimedia Commons, https://en.m.wikipedia.org/wiki/File:Sweep_cultivator_on_the_back_of_a_John_Deere_5220_tractor.jpg.

Disc cultivator for soil preparation. Photo by Gage LaPierre.

multiple seasons, helps reduce weeds in the seed bank. For this technique to be effective, both soil conditions and equipment must be appropriate. During cultivation soil moisture conditions should not be extremely dry or wet. Extremely dry conditions will make the soil difficult to cultivate, whereas wet soil may lead to rutting and loss of soil structure.

A wide variety of implements can be used for this shallow cultivation method, but those that limit soil disturbance to a 2–3-inch depth are preferred. Various tractor-tractor implements can be used, each suited to different soil conditions and site histories. A **rototiller cultivator** is ideal for creating a uniform site condition; it excels at tilling loose soil but is less effective for breaking new ground. For previously cultivated areas, the **sweep cultivator** offers a minimally invasive approach, uprooting or severing young weeds without excessive soil disturbance. When dealing with basic soil cultivation or unbroken ground, the **disc cultivator** is a good choice because it breaks up large soil clods, incorporates vegetation residue, helps somewhat with leveling the ground, and controls weeds by cutting and mixing them into the soil, ultimately preparing a suitable seedbed.

While these tractor-mounted implements are efficient for larger sites, homeowners preparing small meadows can achieve similar results using

Shallow cultivation using a rear-tine rototiller. Photo by Gage LaPierre.

A bare site following shallow cultivation using a rear-tine rototiller. Photo by Gage LaPierre.

walk-behind or hand-operated equipment. For tilling, a walk-behind rototiller or even a garden fork can loosen soil effectively and break ground in a similar fashion as a disc cultivator or tractor-drawn rototiller. Smaller hand cultivators, such as stirrup or Dutch hoes, can replicate the sweep cultivator's shallow soil disturbance action, killing new weeds while flushing out weed seeds from the soil seed bank to the soil surface. These tools allow homeowners to conduct the repeated cultivation technique at a small scale with minimal cost.

STEPS FOR REPEATED CULTIVATION

1. **Spring:** Prior to starting this method, mow the site and clear it of debris. If the site contains a high coverage of perennial weeds or turfgrass, then it may be worthwhile to lightly disc the site. This will help remove deeper-rooted species and chop up debris. Be sure to disc the site when the soil moisture is not too dry (making it hard for the implement to cultivate) or wet (such that a tractor-drawn implement might sink, soil might stick to it, or soil ruts and clods might form). Test out cultivating equipment on a small area to evaluate its effectiveness.

Weeds germinating after first cultivation. Photo by Gage LaPierre.

2. **Early summer:** Shallow cultivation of the site can begin two weeks after initial cultivation or mowing. Select cultivation equipment based on site conditions and set it to a shallow depth of less than 3 inches. Use a roller, cultipacker, or pull-behind drag to smooth out the soil surface post-cultivation; this should cause weed seeds to germinate more quickly.
3. **Summer:** Repeat this process every two to three weeks over the time span of an entire growing season (at least 14 weeks). Weeds should not be allowed to reseed or reach heights greater than 4 inches tall.
4. **Fall:** After one full summer season of repeat cultivation, assess the site for weed pressure. If weed pressure is low, then installation of plant materials and seed can occur. If weed pressure remains high, then we recommend repeating this process again during the following spring and summer seasons.

TIPS FOR REPEATED CULTIVATION

- Sweep cultivators, spiked-tooth cultivators, and rear-tine rototillers are great implements to use for shallow cultivation.
- Discing sod-forming grasses (such as bermudagrass, centipedegrass, bahiagrass, and zoysiagrass) may cause them to spread across the site. To

deter regrowth, apply herbicide several weeks prior to discing. Multiple passes with a disc will be required to reduce the abundance of perennial species on a site prior to starting repetitive shallow cultivation. If the site is dominated by sod-forming grasses, consider using deep sod removal or repeated herbicide application.
- Thoroughly review a site to ensure no debris that will damage cultivating equipment is present (such as rocks, concrete, or metal).

Soil Inversion

Soil inversion works by using a moldboard plow or turn plow to flip (invert) the soil. This technique is typically used on drier upland soils as opposed to wet soils. Notably, soil inversion has been used extensively in the United Kingdom for successful ecological restoration and meadow installations. In Florida there are limited studies testing soil inversion, with the few attempted being unsuccessful. This discrepancy is attributed to improper implementation, highlighting the need for careful planning and expert execution. Therefore, while soil inversion offers significant potential for meadow establishment, it must be approached with caution and expertise to ensure successful outcomes. In 2024 soil inversion was tested alongside other site preparation techniques at the UF/IFAS Plant Science and Educational Unit as part of a project funded by the Florida Wildflower Foundation. The project yielded significantly reduced weed pressure from the weed seed bank at three and six months post-treatment but was not as successful along these metrics as compared to a preemergent treatment or fill-sand method.

The potential advantage of soil inversion is that it can achieve several key objectives in site preparation:

1. It buries existing vegetation, including live weeds, but primarily weed seeds, effectively eliminating them from the soil surface. This creates a relatively clean seedbed, minimizing competition for newly planted meadow species.
2. Soil inversion brings the subsoil to the surface. In Florida, the subsoil is generally less fertile and often lower in organic matter than the topsoil. This is thought to be particularly beneficial for establishing native meadow species, because many are adapted to thrive in nutrient-poor conditions.

3. By reducing soil fertility, soil inversion also creates an environment that is less hospitable to aggressive, nonnative weeds, which generally prefer nutrient-rich soils.
4. The act of inverting the soil also disrupts the soil structure, which on some severely disturbed sites can help to break up compacted layers and improve aeration.

Importantly, sites with sod-forming grasses require pretreatment with herbicides to effectively eliminate these species before soil inversion. The depth of inversion is also paramount for success; a deep inversion, ideally 14–16 inches but at least exceeding 12 inches, ensures thorough burial of the existing weed seed bank and adequate exposure of the subsoil. Achieving this depth requires a powerful tractor (more than 100 horsepower) and proper operation of the moldboard plow. Incomplete turnover of the soil will result in weeds proliferating along the plow furrows, negating the intended benefit. We highly recommend you contract experienced professionals who understand these implements and their proper use.

For residential homeowners aiming to establish a clean seedbed through soil inversion, the traditional moldboard plow is usually impractical for typical yard sizes. A more feasible and manageable approach involves using a deep broad fork or spade, with a blade length exceeding 12 inches, to manually turn over the top layer of soil. While this method requires significant physical exertion, it can effectively bury existing vegetation and seeds, creating a fresh, relatively weed-free surface for planting. To replicate the benefits of a moldboard plow, it's crucial to invert the soil to a depth of at least 12 inches and to ensure that the subsoil is brought to the surface. Following the inversion, using a sod roller can help to compact and level the soil, creating a more uniform planting surface. However, be aware that this method is not suitable for sites with deep-rooted sods, such as bermudagrass. The dense root system of these grasses will impede soil turnover and, in fact, inversion may further spread the sod as the roots are disturbed and fragmented. Therefore, careful consideration of existing vegetation is essential before attempting soil inversion. On sites with deep-rooted sods, alternative methods such as herbicide application or solarization may be more effective.

Moldboard plow in use. Photo by Gage LaPierre.

Moldboard plow in use, showing how it turns the soil. Photo by Gage LaPierre.

Experimental plots after plowing and rolling. Photo by Gage LaPierre.

STEPS FOR SOIL INVERSION

1. **Late summer:** Prior to starting this method, mow the site and clear it of debris. If the site contains sod-forming grasses, then apply preliminary herbicide treatments to kill off these species.
2. **Early fall:** Test the moldboard plow on an area nearby. Adjust the depth of the plow based on the soil series present on the site. Avoid plowing to a depth that brings clay subsoil to the surface. Once the moldboard plow is adjusted properly, invert the soil. This is best performed when the soil is neither too dry nor too wet. A depth of at least 12 inches is recommended, especially on sites with an agricultural history due to the extensive weed seed banks that could be present.
3. **Late fall:** Wait at least two weeks, then lightly disc the site to break up soil clods and kill any potential weed germinates. Follow soon after with seeding and plant material installation.

TIPS FOR SOIL INVERSION

- Do **not** perform two passes or deep till/disc the area after using the moldboard plow or spade, as doing so can potentially bring up new weed seeds.
- Finding moldboard plows can be difficult; check rural feed- and tractor-supply stores for short-term rentals from local farmers or dealers. On small sites, a garden tractor with a miniature turn plow may be capable of mimicking the larger moldboard plow. However, be sure that the turn plow can achieve a sufficient depth.
- If you are considering using a spade to mimic a moldboard plow on a small site, do a test on a small area first to confirm that the physical labor demands are realistic and an adequate depth is achievable.
- Be careful not to damage the moldboard in areas where large rocks or boulders are common. A disc plow may perform better in such areas.

Herbicide Application

Correct use and timely application of herbicides can be an extremely cheap and effective site-preparation method. It is the method most land managers prefer in ecological restoration efforts, especially when nonnative or invasive grasses are present. Multiple applications throughout a season are

top Spot-spraying weeds post-installation. Photo by Nancy Bissett, The Natives.
bottom Pasture site sprayed with glyphosate. Photo by Gage LaPierre.

typically required to create a clean seedbed for installation. For this method to be effective, selecting an appropriate herbicide and correctly timing its application are critical.

The use of herbicides, especially glyphosate (Roundup), has been subject to considerable public debate and legal challenges concerning its human health impacts. The controversy stems from concerns about potential long-term health effects on users and the precautionary principle. Extensive scientific research and regulatory assessments, including those by agencies like the Environmental Protection Agency, have overwhelmingly concluded that glyphosate is safe for use when applied according to label instructions. Safe use, however, relies on responsible use, including wearing proper personal protective equipment and adhering to recommended application rates. When used responsibly, modern herbicides, including glyphosate, can be a very valuable tool for effective site preparation.

OVERVIEW OF HERBICIDES AND BROADCAST APPLICATION

Herbicides are divided into two categories, preemergent and post-emergent. **Preemergent herbicides** kill the seeds of weeds as they germinate, whereas **post-emergent herbicides** kill existing plants. Herbicides can also be classified as **selective**, those that kill only certain plants, or **nonselective**, those that are designed to kill all plants present. Selective herbicides are often separated into those that kill broadleaf plants versus grasses. In most cases for meadow construction, a nonselective herbicide such as synthetically produced glyphosate, imazapyr, or imazapic will suffice for site preparation. Bear in mind that imazapyr can have up to a five-month residual time on sites so consider the installation time frame carefully before use. Organic (nonsynthetic) herbicides are also an option for site preparation. However, most organic herbicides, such as those with acetic acid or citrus oils, are only effective on smaller seedlings (less than 2 in.). Prior to purchasing or applying any herbicide, be sure to read the label and educate yourself on the chemical being used. Not following label requirements may result in failure of a project and even irreversible harm to the site, not to mention potential health risks. Hire a trained and certified consultant or visit a local University of Forida County Extension office for herbicide recommendations or information on how to use herbicides and pesticides safely and effectively.

APPROACHES TO HERBICIDE SITE PREPARATION

There are many different approaches to using herbicides for site preparation. This is due primarily to the range of different herbicides available. In most cases weigh the current composition of the site and examine its seed bank when choosing a strategy. Importantly, also remember that you can use herbicides, either pre- or post-emergent, in combination with other site preparation approaches, as described below.

Example Approach: Sod Removal, Then Post-Emergent, Then Preemergent

One combinational approach that is effective for smallish sites is (1) employ a sod-removal method; (2) wait for seedlings to emerge then apply a post-emergent herbicide (such as glyphosate); and (3) directly following plug installation (no direct seeding), apply a preemergent herbicide (such as prodiamine or pendimethalin) to limit emergence of weed seeds.

Example Approach: Post-Emergent, Then Soil Inversion, Then Preemergent
Another combinational approach good for larger sites is (1) apply a post-emergent herbicide to kill off perennial weeds; (2) follow this a month later with a moldboard plow treatment; (3) directly follow this with an application of a preemergent herbicide; and (4) install containerized plant materials, or alternatively wait the designated period until the preemergent wears off, then proceed to direct seed the site.

Example Approach: Post-Emergent, Then Cultivation, Then Pre-and Post-Emergent
One combinational approach popular among professionals and used across a wide range of site sizes is (1) apply post-emergent herbicide; (2) a few weeks later, cultivate the site; (3) two weeks following cultivation, apply imazapic herbicide, which acts as a pre- and post-emergent; and (4) directly following this treatment, direct seed the site using a carefully selected seed mix that features native species resistant to the preemergent effects of imazapic (see product label for those species).

Other Approaches
There are more combinational approaches and, indeed, many that may not yet be discovered or widely used in Florida. On a large-scale project it is worth consulting with an experienced professional, especially one who has worked chiefly on meadows or restoration projects and has a reputable portfolio available.

BROADCAST HERBICIDE STEPS

1. **Late spring:** Prior to starting this method, mow sites with existing sod and remove the residue, or for previously tilled sites, clear the site of surface debris.
2. **Summer:** Broadcast spray herbicide according to the label instructions and repeat as needed throughout the summer and early fall if label guidelines permit. Applications are most effective when weeds are actively growing, so avoid spraying during droughts. Follow the label recommendations for timing the application relative to expected rainfall. Rainfall too soon after application can reduce herbicide effectiveness.

Large-scale herbicide spraying prior to meadow installation.
Photo by Nancy Bissett, The Natives.

3. **Mid-late fall/winter/spring:** If the site is free of weeds, then proceed to plant seed or install plant material during the planting window recommended for your area of the state. Do **not** till the soil as this can bring new weed seeds to the surface.

BROADCAST HERBICIDE TIPS

- On smaller sites such as residential yards (less than ⅓ acre), a simple pump or backpack sprayer can effectively cover the area. Larger sites are more effectively covered via a tractor-mounted or ATV-mounted boom sprayer.
- Agricultural dyes work well to show areas sprayed or missed by the applicator.
- Herbicide concentrate is almost always more affordable than premixed solutions.

- Preemergent herbicides such as prodiamine, pendimethalin, and dithiopyr are very effective at reducing weed seedling emergence. However, the residual times may last up to five months; also be very thoughtful about their use if you plan to establish the meadow by seed.
- Know your weeds! A selective herbicide may not be effective on a certain type of plants (for example, forbs versus grasses versus sedges), and perennial weeds generally require stronger concentrations or specific types of herbicides and/or repeated application to reduce abundance.

Sod Removal

This method essentially involves removing existing sod on a site to expose bare soil that is ideally free of weed seeds. A **sod cutter** is a machine with a blade that slices the ground beneath the sod, making it easy to roll up and remove. This method is extremely fast and effective on smallish sites (less

left Sod cutter slicing and removing turfgrass for meadow site preparation. Photo by Gage LaPierre.

right The sod cutter removal method is helpful for preparing this small site for meadow installation. Photo by Ross Barreto.

than ⅓ acre), such as residential yards covered in turfgrass. Sites featuring bermudagrass will require additional treatment because bermudagrass commonly has rhizomes that extend deeper into the soil than sod cutters can reach. Rhizomes left in the soil can resprout. Because a site can still contain an extensive weed seed bank following sod removal, it is worth considering a second site preparation treatment. Applying a preemergent herbicide or solarizing the site can prevent weeds from later overtaking the meadow.

STEPS FOR SOD REMOVAL

1. **Midsummer:** Mow the site to a short height and thoroughly check for debris and irrigation lines, which can damage the sod cutter.
2. **Late summer:** Rent a sod cutter and proceed to remove the turfgrass on the site. Begin by testing the sod cutter in a nearby area and ensure it is set on the maximum depth. Once finished, review the site for any remaining roots and remove them by hand or rake. Monitor the site for at least one month for a return of turfgrass or an influx of weeds. As weeds come in, remove them via herbicide, by flame torch, or by hand. Do not allow weeds to grow taller than 2 inches in height. Avoid tilling or digging in the site, which can bring new weed seeds up to the surface.
3. **Early fall:** If the site is largely free of weed seedlings and turf (at least 95 percent bare ground), then proceed to install plant material. If the site does not meet this criterion, postpone planting until late fall or early winter and continue the weeding treatments described in step 2.

TIPS FOR SOD REMOVAL

- This method does **not** work on sites featuring more than 25 percent coverage of bermudagrass because the rhizomes commonly exceed the maximum depth of sod cutters.
- Pay close attention on sites with mixed species of turfgrass to ensure removal is effective with all types.
- Bahiagrass (*Paspalum notatum*), zoysia (*Zoysia* spp.), centipedegrass (*Eremochloa ophiuroides*), and St. Augustine grass (*Stenotaphrum secundatum*) can be removed effectively with this method.
- St. Augustine and zoysia turfgrass, if cut properly, may be capable of being recycled or sold.
- Many local equipment outfitters rent sod cutters by the day at affordable prices.

top Fill sand helps to cover the weed seed bank. Photo by Gage LaPierre.

bottom left Fill sand has been spread across this site to cover weed seeds. Photo by Gage LaPierre.

bottom right A thick layer of fill sand has been evenly spread across a site to cover the weed seed bank. Photo by Gage LaPierre.

- If the sod is thick and healthy, try removing it from the site by rolling it up after cutting instead of chopping it into pieces. Use a strong metal pipe to pick up the sod rolls from the site.

Covering the Weed Seed Bank

The basic idea of this site treatment process is to suppress the weed seed bank (reservoir of dormant weed seeds within the soil) on the site by covering it with a layer of mined fill sand thick enough to inhibit germination. This method has limited real-world examples to draw from but has been reported as successful in several tangential examples. This method will not work by itself but can support other site preparation treatments. It will also be ineffective if the site has invasive species or nonnative grasses. This site preparation method is well suited for smaller residential sites where turfgrass has been removed with a sod cutter or eliminated with herbicide. Mined fill sand can be purchased and delivered to a site in bulk quantities at a low cost ($250–$400 per 18 cu. yd. in 2024). Apply the fill sand at a minimum thickness of 1.5 inches to adequately suppress weed seeds.

STEPS FOR COVERING THE WEED SEED BANK

1. **Midsummer:** Mow the site to a short height, then thoroughly check for and remove debris. Use herbicide on the site or remove turfgrass with a sod cutter and follow with light cultivation or a preemergent herbicide.
2. **Late summer:** Review weed pressure on the site. Apply herbicide to emergent weeds (preferred) or remove them by hand or light cultivation.
3. **Early fall:** Review weed pressure on the site again. Kill emergent weeds using your preferred method, then proceed to spread mined fill sand across the site using a commercial drop spreader or tractor loader, or by hand if feasible. Be careful not to stir the sand into the ground, which exposes it to or mixes it into the site's soil; doing so will expose weed seeds to the surface. Immediately after spreading the sand, the site can be direct seeded or planted.

TIPS FOR COVERING THE WEED SEED BANK

- Consult reviews of mined fill sand sources to ensure cleanliness and purity of the material.

- Pay close attention to sites with mixed species of turfgrass and avoid using this method on sites with greater than 25 percent coverage of bermudagrass.
- Avoid using fill sand with high amounts of clay as this can inhibit seeded species germination.
- Use of a commercial drop spreader is highly recommended to ensure that the material is spread across the site uniformly.
- Do not roll the site after spreading fill sand because this can bring weed seeds closer to the soil surface.

Dealing with Improved Pastures and Abandoned Agricultural Sites

Improved pastures are areas that were cleared and planted in perennial forage. Improved pastures and abandoned agricultural fields cover more than 20 percent of the total land area of Florida.[2] Most improved pastures in Florida are planted in nonnative, sod-forming grasses, primarily bahiagrass (*Paspalum notatum*) or bermudagrass (*Cynodon dactylon*) but sometimes

Improved pastures are potential sites for meadow installation in Florida. Photo by Gage LaPierre.

Abandoned agricultural lands are also potential sites for meadow installation in Florida. Photo by Gage LaPierre.

also unintentionally planted centipedegrass (*Eremochloa ophiuroides*). It is essential to remove these species prior to meadow installation to minimize their competition with planted material. However, these grasses, especially bermudagrass, can be extremely difficult and costly to remove. Herbicide treatment methods are the most successful and efficient means of eliminating bahiagrass and bermudagrass from a project site; sod removal is not effective. However, because of its deep root system, bermudagrass may require up to two years of treatment for removal on a large site, and unfortunately even then it may not be totally removed. Ultimately, total removal may not be practical or necessary to achieve the goals of many projects. Competition pressure from bermudagrass on established plants and on meadow community composition is not well studied. Research is still ongoing to determine what level of removal is needed for the long-term success of a project. We advise you work cautiously on such sites and strongly recommend you seek professional advice.

Bahiagrass (*Paspalum notatum*) is an invasive sod-forming grass common in Florida pastures. Photo by Mark Marathon, 2012. Creative Commons license via Wikimedia Commons, https://commons.wikimedia.org/wiki/File:Eremochloa_bimaculata_stolon.jpg.

Centipedegrass (*Eremochloa ophiuroides*) is an invasive sod-forming grass common in Florida pastures. Photo by cwarneke, 2017. Creative Commons license via iNaturalist, https://www.inaturalist.org/observations/7267036.

Bermudagrass (*Cynodon dactylon*) is an invasive sod-forming grass common in Florida pastures. Photo by leaf0605, 2024. Creative Commons license via iNaturalist, https://www.inaturalist.org/observations/236222046.

above left Containerized plant being installed. Photo by Nancy Bissett, The Natives.

above right Installing containerized plants in a meadow can supplement direct seeding efforts. Photo by Gage LaPierre.

Plant Material Installation

Once the site has been properly prepared, then plant material installation may begin. To minimize the potential for stand failure, reduce costs, and achieve species composition goals in a relatively short time frame, we recommend a combination of direct seeding and planting of containerized plant materials for Florida meadows.

Determining the best sequence and timing of planting relative to seeding is still a very active area of research in Florida. However, in general, if you install containerized plants after seeding, do so before seeds germinate (within 48 hours) and avoid burying seeds. If you choose to plant containerized materials prior to seeding, be careful to leave enough space to allow for rolling or pressing seeds into the soil while avoiding harm to planted material.

opposite Plugs of plants ready for installation. Photo by Gage LaPierre.

left A Grasslander seeder. These are usually tractor-drawn. Photo by Gage LaPierre.
right Grasslander seeder loaded with seed for sorting. Photo by Nancy Bissett, The Natives.

Direct Seeding

Successful direct seeding can be challenging even for professionals with extensive experience. Many factors can negatively impact the germination and establishment of seedlings. These include natural factors that are often outside of human control, such as seed predation, seed drift from wind or flooding, and seedling mortality due to drought or intense solar radiation. You can alleviate some of these issues by controlling other factors like the timing, quantity, seed mix composition, and method of seeding. Any of these factors alone can greatly influence the success or failure of a direct seeding effort.

TIMING OF SEEDING

In Florida, the best time to seed a meadow is during the fall and winter months (September–February), and seeding outside of this period is not recommended. The reason is because species that flower in the spring typically germinate more readily from September to November, whereas summer- to fall-flowering species tend to germinate from November to February. Cold weather fronts during this time bring rain and colder conditions that many native grasses and wildflowers favor for germination in the wild. Some grasses and wildflowers even require cold stratification periods for seed germination to occur. Bear in mind that the start of fall and winter weather conditions varies each year depending on factors such as cold snaps or tropical warm fronts. These weather conditions can also vary drastically along Florida's latitudinal gradient. Hence, in the northern Panhandle spring-flowering species may be best sown in early September while in South Florida the same species may be best seeded in late November.

SEEDING RATE

Seeding rate recommendations are based on the expectation that if you plant a given number of seeds, the desired number of plants will grow in the area planted. The current recommendation for meadow planting is 40–60 viable seeds per square foot when using germination-tested seed (see "Direct Seeding Methods" on the rationale for the range).

Additionally, many native species have appendages that assist with natural dispersal (such as pappus on *Liatris* spp. and awns on *Aristida* spp.) that either cannot or should not be completely removed. Consequently, variable and often significant levels of inert plant material (non-seed) will be present in many native seed lots. Thus, with most native plants it is equally important to know the amount of inert material present in a seed lot.

The best way to guarantee achieving the desired seeding rate is to use seed lots that have been tested by a testing agency and have tags that show the information necessary to calculate **pure live seed** (PLS). PLS represents the percentage of viable seeds of the desired species in the bag. This is not printed on the seed tag but can be calculated (see below). When you purchase seed on a PLS basis, you can be sure you are paying for viable seeds of the particular species. For a given PLS seeding rate, the equivalent pound-per-acre rate can vary greatly between seed lots. This may mean that the seed lot with the cheapest per-pound cost turns out to be more expensive overall if the germination rate is low or it contains a lot of inert material.

When using wild-harvested material or field-grown seed that has not been tested, the recommended seeding rate is 80–120 seeds per square foot. The recommended seeding rate for these types of seed is higher than the PLS rate due to the potential factors that could result in germination failure which have not been accounted for by cleaning and testing. Before using either wild-harvested or field-grown seed, perform a basic germination test and use this information to adjust the seeding mixture rate to achieve your desired objectives.

SPECIES SELECTION IN SEED MIXTURES

Seed mixtures typically contain around a dozen species, but some have been known to include more than 70 species. High-diversity seed mixtures tend to cost more but are thought to increase the likelihood of finding plants that match the site well enough to establish and persist. As a general rule of thumb, seed mixtures with a one-to-one graminoid-to-forb ratio produce quality results.[3] Higher forb ratios may lead to more bare ground prone to weed infestation, whereas higher grass ratios are likely to lead to lower overall plant diversity. Seeding mixtures should contain a low to moderate quantity of ruderal grasses and short-lived forbs such as lovegrasses (*Eragrostis* spp.) and tickseeds (*Coreopsis* spp.). These species are known to establish readily by seed and fill in space that would otherwise become occupied by weeds. See appendix B for examples of seed mixture recipes.

Calculating Pure Live Seed

PLS is determined by multiplying the percentage of pure seeds by the percentage of seeds that germinate. (The percentage purity and percentage germination are listed on the seed tag.)

For example, assume a bag of switchgrass seed has 70 percent germination and 80 percent purity:

$$70 \text{ percent germination} \times 80 \text{ percent purity} \div 100 = 56.$$

In other words, only 56 percent of the material in the bag is viable seed. Then divide the recommended seeding rate by the PLS to determine the amount of seeds needed:

$$10 \text{ lb./acre} \div 0.56 = 17.9 \text{ lb./acre}$$

Thus, you will need to plant 17.9 pounds per acre of the switchgrass seed to provide 10 pounds per acre of PLS.

Measuring Seed Viability of Wild-Harvested Seed

Seed harvested from the wild may have variable seed viability and should always be tested prior to widespread use. Seed viability can be measured through a variety of techniques that include a germination test, seed staining, and hyperspectral imaging. **Seed staining** involves the use of dyes, typically tetrazolium, to stain seeds with viable embryos, whereas **hyperspectral im-**

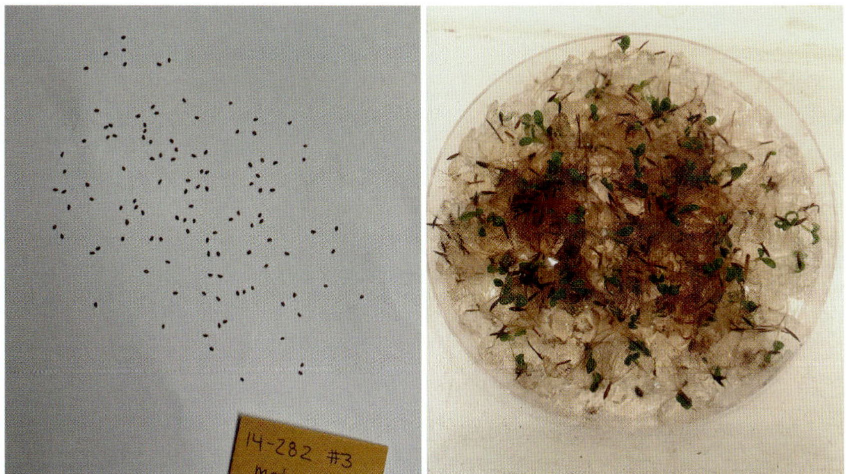

left A batch of seeds such as these ones displayed at the Florida Brooksville Plant Materials Center can be sent off to a private company for testing in batches or tested in-house via a bench germination test. Photo by Gage LaPierre.

right A bench germination test involves placing a given number of seeds in a petri dish filled with sterile agar and allowing them to germinate over a given time. Photo by Gage LaPierre.

aging uses X-rays to visualize viable seed embryos. Both methods are complex and require specialized training. A **germination test** is the simplest and cheapest method for testing seed viability. One easy method for performing a germination test is to place a given number of seeds (typically 25 seeds) into a petri dish containing moist germination paper or a polymer or agar solution. Maintain the seeds in ambient conditions for the next 30 days and monitor for germination. After 30 days a germination rate can be calculated.

Availability of Seed Mixtures

In many cases, it may be more reliable to transplant long-lived perennial species via container rather than to pay for seed mixtures. This is particularly true if the seed is from an out-of-state source. The limited availability of seeds in the marketplace is a major issue facing the native plant industry, as demand often outpaces the supply of seeds in the marketplace. It is wise to purchase seeds far in advance of the installation date to procure your desired species or locate good alternatives.

DIRECT SEEDING METHODS

There are three ways to direct seed meadows: broadcasting, drilling, and hydroseeding. There are significant pros and cons associated with each method in terms of cost, project size, location, and equipment requirements.

Broadcast Seeding

Broadcast seeding is by far the simplest, cheapest, and most common method for seeding a meadow, especially on small sites. However, broadcast seeding is not recommended for steep slopes or roadside shoulders because of the potential for soil erosion and the seeds blowing or washing away.

Rotary spreaders (equipment used to spread crop seeds, fertilizer, or various types of granular products) can be used to broadcast all types of native seed over relatively large areas. On smaller sites, seeds can be spread by hand, a handheld seed spreader, or a pushcart whirlybird.

Standard broadcast spreaders, particularly handheld types, have trouble with seed mixes that contain a lot of fluffy seed or trashy inert material. Handheld or ATV-mounted broadcast spreaders that have specialized gears, or fingers, that "pick" through fluffy seeds are a better choice when that type of seed is being used.

Regardless of the type of broadcast planting method, mix the seed with some sort of diluting agent, such as rice hulls, plain clay kitty litter, or even

left A hand-spread seed mixture appropriate for smaller sites. Photo by Gage LaPierre.

middle The hand-spread seed mixture will spread more evenly when mixed with a diluting agent like sand. Photo by Gage LaPierre.

right This photograph shows (dark-colored) seeds evenly mixed throughout (lighter-colored) sand used as a diluting agent. Photo by Gage LaPierre.

dry sand, to more evenly distribute the seed across the site. A good starting point for diluting your seed mix is a one-to-one ratio (by weight) of filler to seed. However, this may need to be adjusted upward to achieve the correct seeding rate and ensure even distribution, especially when dealing with very small or fluffy seeds or using equipment that requires a certain bulk weight to function properly.

Prior to broadcast seeding, run a cultipacker or roller over the area to firm the seedbed and ensure good seed-to-soil contact. After broadcasting

top left A cultipacker improves seed-soil contact. Photo by Nancy Bissett, The Natives.

top right This ground shows an even distribution of broadcast seeds. Photo by Nancy Bissett, The Natives.

bottom Seeds can also be broadcast across a site using a hay blower. Photo by Nancy Bissett, The Natives.

Native seed drills such as the Grasslander or similar rangeland seed drills are useful when seed is being applied directly over a large site. Photo by Nancy Bissett, The Natives.

the seed, we recommend running over the area with a light disc, roller, or cultipacker to help ensure seeds make good contact with the soil and are not subject to being blown or washed away easily. On a small site, spreading a thin layer of pine needles evenly across the ground will help reduce soil erosion as well as seed loss to wind and flooding.

Seed Drilling

Slick-seeded native species can be planted using standard agricultural-type drills. A seed drill works by placing seeds in one or more seed hoppers from which the seed falls out. This allows different-sized seeds to be planted at different rates or even depths, if necessary. As the drill moves across the site, the seeds from the different hoppers are dropped through tubes either onto the soil surface or into furrows created with discs preset at given depths. Typically the depth would be no greater than ¼ inch with a range of ⅛ to ½ inch. Press wheels behind the machine then press and firm the soil over the seed. If fluffy-seeded species like bluestems or blazing stars are in the mix, or a lot of inert material (stems, leaves, and so on) is present, you will need a native seed drill or range planter for direct seeding on a large site. In a **native**

drill or range planter, at least one of the hoppers has specialized gears or fingers that can "pick" through the fluffy seed mixture without it clumping.

Cultipacking or rolling after planting with a drill is highly recommended but not required. Disadvantages of drilling are that the equipment is generally only useful on a relatively large area, and range drills and planters are not readily available for lease in Florida. However, custom planting with such equipment can be contracted.

Hydroseeding

Hydroseeding is commonly used for resodding roadsides and construction sites that are steeply sloped. This method essentially involves creating a mixture of seed, mulch, and water and blasting the resulting slurry onto the site via high-pressure pumps. When hydroseeding native species, it is critical not to overapply mulch as it can smother the seeds beneath it. Around 2,000 pounds per acre of mulch, applied in a two-step process, is recommended for hydroseeding with natives. First, apply a mixture of seeds with half the desired amount of mulch to the site, then spray the remainder of the mulch on top. Do not add tackifiers or fertilizers to the slurry: Tackifiers will inhibit the germination of certain species of wildflowers, while fertilizers will favor the growth of weeds. Overall, hydroseeding is a very expensive means of direct seeding due to the specialized equipment and experienced personnel needed. A significant amount of water may also be required when covering a large site. Ground cover restoration research in Florida found that on flat terrain this method was no more effective than broadcast seeding or seed drilling. Therefore, unless the site is steeply sloped, hydroseeding is not recommended for installation of meadows in Florida.

Containerized Planting Materials

Again, it is strongly recommended that some level of containerized plant materials be used. When installing containerized plants, the biggest factors to consider are spacing and size of the species.

In terms of size, containerized material generally comes in either standard nursery trade pot sizes (½ gal., 1 gal., 3 gal. and so on) or as plugs that are about 2 to 4 inches wide and range from 5 to 12 inches deep. Plugs are generally substantially cheaper to purchase in mass and to install compared to potted material. Potted plants are much costlier than plugs to purchase and install, but ideally, they will fill in and establish on sites more quickly. Note that potted plants can at times be more susceptible to injury and stress mortality than plug material and can create substantial soil disturbance during installation. Some species are only available as potted plants and vice versa, so a mix of both plug and potted materials is common.

Spacing should be based on anticipated mature plant size, but a minimum distance of 7 inches apart is recommended to avoid self-competition, while the maximum distance in most cases should be no farther than 30 inches so that plants fill in effectively during the first growing season. Larger spaces can be left if the site has been or will be seeded at sufficient densities.

TIPS FOR INSTALLING CONTAINERIZED PLANTS

- Plugs are best planted using a dibble; flat dibble bars work well on sites with compacted clay soils, whereas cone-shaped dibble bars perform best in sandy or loose soils.
- On a large site an earth auger may prove handy for installing potted materials.
- Applying water polymers, such as Soil Moist or TeraSorb, to the roots of plants may reduce plant mortality and improve establishment on xeric sites. However, be careful to avoid use in clayey soils and do not overapply because doing either may result in poorer growth or establishment.[4]
- Always be sure volunteers and staff assisting with the project are edu-

left Planting a meadow with containerized plants. Photo by Nick Freeman, Wakka Pilatka, LLC.

right People planting a meadow with containerized plants. Photo by Gage LaPierre.

left A flat bar dibble is useful for planting in compacted soils. Photo by Randy Fernandez.

right An earth auger is useful for planting larger materials such as potted plants. Photo by Joe Reams, Southern Habitats.

cated in proper planting techniques such as correct planting depth, appropriate hole size, gentle root handling, firm soil contact, and spacing.

- Do not use leaf mulch, wood chips, sawdust, or other wood-based mulches, because they may physically obstruct the meadow seed from germinating and remove nitrogen from the soil. Any type of mulch, other than a light layer of pine straw, should be used sparingly around transplants, particularly if direct seeding is planned between transplants. Again, excessive mulch can physically prevent meadow seeds from germinating.

Post-Planting Care

Prolonged droughts and flooding are typically the biggest issues directly following an installation. Drought can stifle germination and cause plant mortality, which can only be remedied by irrigating the site. If practical, daily irrigation for the first two weeks is recommended. Provide at least ¼ inch of water, but higher amounts, up to 1 inch may be needed on extremely well-drained soils like Alachua County's Candler series. It is best to irrigate in the early morning to reduce evaporation during midday heat. Flooding, on the other hand, can cause issues with erosion and seed translocation. The best way to mitigate this issue is to time planting to avoid predicted extreme precipitation events.

four

MANAGING A MEADOW

All meadows need management to stay healthy, attractive, and floristically diverse. Conceptually, meadows are a novel plant community created by humans. Consequently, meadow management is inherently different than maintenance of typical urban landscapes. In particular, meadows are maintained as a community, rather than as a collection of plants that require individualized maintenance. Over the long run, this translates into far less maintenance when compared to a typical landscaped or mowed space.

The type of meadow management protocol selected is largely determined by the desired goals and maintenance methods available. The goals and purpose of meadow management should always center around controlling weedy species in favor of more desirable species. In some cases, goals may also include minor elements of regulating the timing of blooms, maintaining vibrancy, or retaining habitat qualities for certain wildlife species. Again, setting metrics for meadow composition and structure is important for creating management strategies and should be done in advance of installation (see chapter 2). Keep in mind, too, that expectations can often differ substantially among parties involved in the project; for example, between client versus homeowner or homeowner versus neighbors. Thus, it is worth having conversations early on to keep short-term objectives in line with long-term goals. For example, cutting back a meadow before sets of species flower may seem foolhardy to some in the short term but may be a necessary part of the plan for reducing weed abundance over a longer time frame. All too often, though, pressure for short-term gains wins out over long-term management strategies when managing meadows.

All long-term management strategies that reduce or filter out weedy species involve prescribed burning, mowing, applying herbicide, grazing, and hand-weeding. The seasonal timing and frequency of these treatments can greatly influence the effectiveness of each treatment and outcome. Other compounding factors to consider are the intensity and scale of treatments

opposite Meadow designed by My Florida Meadow Company along a driveway. Photo by Andrea England with My Florida Meadow Company.

A prescribed fire creeping across a residential meadow. Photo by Troy Springer.

The same meadow three months post-burn, showing new growth and regeneration. Photo by Troy Springer.

(that is, their area or size within a meadow). Do not overlook any of these compounding factors when crafting a management strategy. Also, do not be afraid to adjust your strategy if your objectives are not met. **Adaptive management**, as it is known, is a cornerstone in land management and restoration ecology. The basic premise behind this idea is to monitor a site, evaluate effects of past actions, and adjust future actions based on those evaluations.

Management Conceptual Diagram

The "Management Strategy" diagram illustrates a comprehensive approach to developing a successful meadow management strategy. The process begins with clearly defined goals, which may encompass aesthetic, economic, environmental, cultural, and wildlife considerations. These goals, in conjunction with an assessment of existing site conditions—such as the presence of weeds, the success of meadow establishment, and prevailing weather events—inform the selection of appropriate management treatments. These treatments may include fire, mowing, herbicide application, and hand removal. It's crucial

Management Strategy

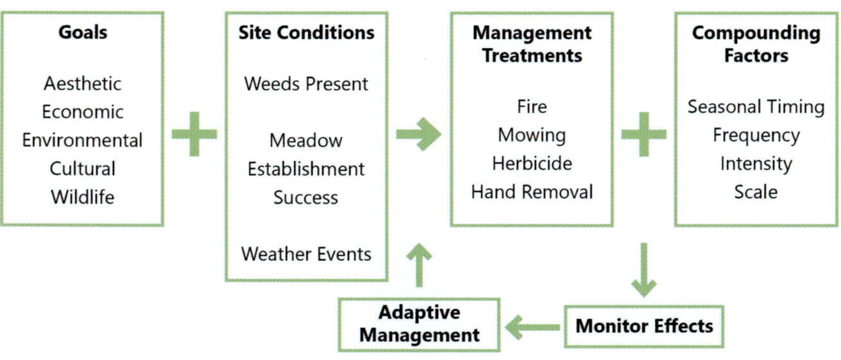

The goals of the meadow installation combined with the site conditions inform the best management strategy. A management strategy boils down to balancing management treatments against compounding factors associated with those treatments. All management strategies should include monitoring the effects of treatments, reviewing those effects, and potentially adapting the management strategy to improve results. Illustration by Isabella Guttuso Browne.

to recognize that the effectiveness of these treatments is also influenced by compounding factors like seasonal timing, frequency of application, intensity, and scale. To ensure the management strategy remains effective and aligned with the initial goals, a robust monitoring program is essential. Carefully observing the effects of the implemented treatments and subsequently monitoring these effects allows for an adaptive management approach. Adjusting the management strategy over time as conditions change will ensure the long-term success and health of the meadow.

Management Timeline Example

The first year they sleep, second year they creep,
third year they lead. —UNKNOWN

In designing a management strategy for a meadow, the general rule of thumb is that management during the first three years will always be more intense than in subsequent years. The primary reason for this is that it takes roughly three years for the seeded or planted perennial component of a meadow to become fully established and fill in the area, although this time frame may be somewhat shorter if containerized plant materials are used for perennial species. The main objectives during the first few years should be to ensure that desired plants become established and that weed coverage remains low. If these objectives are not met, it will be difficult to achieve any of the goals that were identified for the meadow planting.

In designing a management strategy for a meadow, the general rule of thumb is that management during the first three years will always be more intense than in the subsequent years. Illustration by Isabella Guttuso Browne.

Year One Management

The first year following installation is the most maintenance-intensive time for meadow management, primarily because weed pressure tends to be highest.[1] This is because there is the greatest amount of bare ground in the first year for weeds to exploit. Proper site preparation will minimize weed establishment, but even ideal site preparation cannot eliminate all weed pressure. Additionally, some weeds, generally those that are stoloniferous or rhizomatous perennials, are more problematic than others (see appendix C for details). Monitoring the site and identifying weeds promptly is critical to proper management.

For most plantings, mowing treatments should be the primary means for weed control during the first year. During this period mowing should occur whenever the average height of the meadow exceeds 18–24 inches, cutting it down to a height of 7–12 inches.[2] On productive soils this may translate into mowing being necessary four to six times in one year, versus two or three times on less productive soils. Mowing this frequently during the first year ultimately helps control fibrous ruderal species (such as ragweed, dogfennel, and horseweed) that would otherwise outcompete installed plants, especially seedlings. Avoid mowing desirable annuals when they are flowering or developing seeds. Mowing prior to or after these periods will allow annuals to grow back and reseed themselves.

Depending on what weedy species are present, hand removal or spot treatment with a labeled herbicide may be more practical than mowing to control weeds. If any known problematic or invasive species encroaches on the planting, herbicide spot treatment should be the first management action implemented. Keep in mind, though, that regardless of treatments, most sites will suffer from an influx of annual weeds such as crabgrass (*Digitaria* spp.), Florida pusley (*Richardia* spp.), and chamberbitter (*Phyllanthus urinaria*). Keep in mind, too, that annual weeds like these are difficult to control with any treatment due to their rapid growth and reseeding ability. Luckily, these types of weeds generally are not very problematic in meadows as they tend to be outcompeted by desired species in the long run.

Year Two Management

Managing the meadow during year two is much easier if proper management was followed during year one. Desired perennials should be flowering and have a coverage exceeding 50 percent by the spring season of year two. If this is not the case, then the addition of more plant materials in the form of plugs is warranted. Management during this period should ideally consist of one or two mowings for weed control and to stimulate new growth or flowering. One mowing treatment or multiple selective patch mowing treatments should be sufficient if weed pressure is low. If necessary, treat large patches of weeds or individual invasive species with herbicide, mowing, or hand removal.

Year Three Management

By year three desirable species should dominate the site completely with minimal weed pressure and limited management intervention. One mowing treatment targeted for late spring or early summer should be sufficient to maintain meadow composition. At this stage, a prescribed burn may replace mowing on sites that can be safely and effectively burned. If the meadow features trees, then be cautious of flame intensity as intense flames may result in tree damage or mortality.

Management Beyond Year Three

Post year three most meadows will continue to change in composition over time. Longer-lived perennials, for example, will tend to become more dominant over annuals. To encourage large flushes of annuals, avoid mowing or burning prior to seed setting. Various species of shrubs, such as saltbush (*Baccharis halimifolia*), wax myrtle (*Morella cerifera*), and laurel cherry (*Prunus caroliniana*), along with some vines, may find their way into the meadow. These species should not harm long-term composition if active management (mowing, prescribed burning) is continued regularly. The composition of the meadow may also be affected by encroachment of turfgrasses if they are present along the meadow boundaries. Proper upkeep of these borders (such as edging, herbicide) is important in limiting the encroachment of turfgrass. This should be followed by periodic spot-checking for weeds or invasive plants. See appendix C for common weeds and how to treat them.

Keep in mind, too, that as a meadow develops past year three it will continue to change according to the site conditions, climate, and management. For instance, certain species may advance into or retreat from areas of the site due to soil or light conditions. In some cases, mowing or fire regimes may filter some species out of the meadow. For example, early spring burning can kill off some annual species like goldenmane tickseed (*Coreopsis basalis*) or Drummond's phlox (*Phlox drummondii*). Similarly, repeated mowing can reduce the coverage of younger forb populations. Rotating fire and mowing treatments from year to year, or even making slight changes in the seasonality of these practices, can maximize diversity by creating a mosaic of different disturbance patterns. This prevents any single species or group of species from becoming overly dominant, allowing a wider range of plants with varying life cycles and tolerances to thrive.

As our climate continues to change, it makes sense that meadows will change as well. Shifts in temperature, rainfall patterns, and the frequency of extreme weather events such as droughts or floods will directly impact plant communities. However, diverse meadows are thought to have a high capacity to endure these events, offering a degree of resilience that monoculture landscapes lack.[3] This resilience stems from the variety of plant species within a diverse meadow. Different species have various tolerances for environmental stressors; some may be more drought resistant, while others may be better adapted to waterlogged conditions or temperature extremes. This diversity acts as a buffer, ensuring that even if some species are negatively affected, others will persist, maintaining the overall function and structure of the meadow. Periodic monitoring of a site will help in understanding changes to its composition, allowing for adaptive management strategies to ensure long-term health and resilience as needed.

Management Treatments

Ongoing management techniques include mowing, burning (where safe), and herbicide use. Mulching generally is counterproductive in meadows, but limited application of pine straw may be useful.

Mowing

Several types of mowers are available, each with different benefits and drawbacks.

FLAIL MOWER

Flail mowers are thought to be ideal for mowing meadows because they can chop material into fine pieces that tend not to smother plants or open soil

Disc mower. Photo by Christer Folkesson, 2008. Creative Commons license via Wikimedia Commons, https://commons.wikimedia.org/wiki/File:JF_GX_2400_SM_mower_with_conditioner.jpg.

Flail mower. Photo by Alupus, 2010. Creative Commons license via Wikimedia Commons, https://commons.wikimedia.org/wiki/File:Schlegelmulcher.jpg.

Rotary mower. Photo by BulldozerD11, 2009. Creative Commons license via Wikimedia Commons, https://commons.wikimedia.org/wiki/File:Teagle_Dynamo_245_finishing_mower_-_IMG_0423.jpg.

that annuals need for reseeding and solitary bees prefer for nesting. A flail mower also minimizes flying debris and handles rough terrain and overgrown areas with ease, providing a clean cut even on dense brush and weeds. Flail mowers, however, tend to be much more expensive than other mower types and usally require a dedicated walk-behind machine or a tractor to operate.

ROTARY MOWER

A rotary mower is equipped with a large steel blade that spins rapidly beneath a protective metal deck, effectively chopping vegetation. These mowers are the most widely used type in the United States, with many homeowners relying on small, push-style rotary mowers for lawn care. Additionally, rotary mowers are the predominant choice for tractor attachments used in roadside and rural maintenance and by commercial lawn care companies, which often prefer zero-turn rotary mowers for their efficiency and precision. The popularity of rotary mowers stems from several key advantages, including versatility, ease of operation, affordability, and relatively simple maintenance compared to other types of mowers. Moreover, these mowers excel at handling obstacles such as twigs and small rocks, cutting through them more effectively than alternative designs do. Rotary-style mowers have been used effectively in the maintenance of meadows. The biggest issue with using this type of mower in a meadow is that many models feature limited height control, making them impractical for cutting tall vegetation without a high-powered engine. Rotary mowers also have the potential to scalp vegetation or for the mower deck to drag along the ground in uneven areas, both of which can potentially damage both the mower itself and the meadow. Rotary mowers usually also produce significantly more dust than other mower types.

SICKLE BAR MOWER

Sickle bar mowers are a relatively uncommon type of mower in Florida today. They are generally more expensive than rotary mowers and require more maintenance. Small walk-behind sickle bar mowers work well, though, for small urban meadows where tractors are prohibitively large and rotary mowers are incapable of moving in or cutting the material. Sickle bar mowers are also capable of cutting tall material with one pass, which other mower types may struggle to accomplish. They also tend to produce less dust and be better able to cut along angled terrain.

Sickle bar mower. Photo by USDA Media by Lance Cheung, 2016. Creative Commons license via Flickr, https://www.flickr.com/photos/usdagov/27599010092, reproduced with cropping.

STRING MOWER

String mower equipment is common and inexpensive, with most residential homeowners having a small shaft-driven machine. Heavy-duty walk-behind string trimmers are also commonly available and are capable of cutting down fairly tall vegetation. While string trimmers can be used to maintain small meadows, we do not recommend them because they tend to cut messily and nonuniformly. The string heads on this equipment also tend to get bound up in larger grass species.

POWER SCYTHE

A power scythe is commonly used as a hedge trimmer, but it can make a excellent tool for mowing small residential meadows. Articulating models of the power scythe work best because they allow you to cut evenly while standing comfortably. Electric models of this tool are also commonly available, which reduces emissions as well as noise and dust pollution, making it even more attractive in residential settings.

Fire

Fire is inherently dangerous, but when performed safely, a prescribed fire or controlled burn is an extremely effective management tool for maintaining highly aesthetic and diverse meadows. Prescribed fire, when timed

Notes on Mowing

- There is no overall ideal mowing height for meadows, and the goals for the site can affect mowing heights substantially. Mowing heights as low as 2 inches to as high as 15 inches have been reported in meadow case studies as successful for achieving particular goals.
- Mowing below 8 inches will generally discourage woody or fibrous species in a meadow. However, this height may favor nonnative sod-forming grasses and prostrate weeds (such as Florida pusley, bahiagrass, and bermudagrass) if present. Mowing at this height more than twice a year can damage younger desirable forbs.
- Mowing above 15 inches is generally safer for desirable species but may also result in high survivability of weedy perennial forbs (such as ragweed, Spanish needles, horseweed). Greater frequency of mowing may help compensate for higher weed survivability.
- Late-winter and midsummer seasons are the preferred times to mow meadows. Late-winter mowing avoids harming nesting insects, including some pollinators, and helps promote spring wildflowers. Meanwhile, midsummer mowing does not affect spring wildflower stands while promoting fall-flowering species.
- Flail mowers seem to be the best mower option for mowing meadows, as they chop the material finely and feature easy height adjustments. However, rotary, sickle bar, power scythe, disc blades, and string mowers can work effectively as well.
- Always clean mowing equipment thoroughly prior to mowing because weed seeds can stick to the mower and be introduced into the site.
- When mowing landscaping along the borders of a meadow, be sure to face the mower chute away from the meadow to avoid weed seeds being flung and spread into the meadow.
- In small meadows use of a leaf vacuum or cyclone rake following mowing can help remove cut biomass from the site for a cleaner aesthetic.

Controlled burning of a meadow. Photo by Troy Springer.

correctly, offers numerous benefits for meadow management, mimicking natural disturbance regimes that have shaped native ecosystems. By consuming accumulated organic matter, fire lowers nutrient levels and reduces thatch, creating conditions that favor native species adapted to low-nutrient environments. Fire also effectively reduces many nonnative weed seedlings, particularly annuals and short-lived perennials like Spanish needles, crabgrass, and ragweed, reducing the competition facing desirable plants over time. Beyond weed control, fire serves as a powerful cleanup tool, removing dead vegetation and debris and preparing the ground for new growth. Importantly, fire triggers flowering in many native species (for example, silkgrass [*Pityopsis* spp.], wiregrass, and blazing star), which promotes seed production and attracts pollinators that enhance biodiversity. The resulting flush of new growth also provides fresh forage for wildlife and contributes to a vibrant, healthy meadow community. Furthermore, fire creates valuable patches of bare ground, essential for the successful seeding of many native species and the creation of optimal nesting habitats for ground-nesting bees, which rely on these open areas for their survival.

Careful planning and experience are necessary to effectively employ fire as a management tool for meadow maintenance. Prescribed fires should only be attempted by experienced and trained individuals. In Florida, it is illegal to perform a prescribed burn on your property without a permit from the

Florida Forest Service (FFS). Individuals seeking to use prescribed fire as a management tool are encouraged but not required to take the prescribed burner course offered by the FFS. Alternatively, an FFS-certified prescribed burner can be contracted to conduct the prescribed burn. However, burn permits are unlikely to be issued within urban residential areas, so burning is not a practical management option on such sites. It is also worth mentioning that even in areas where burn permits are generally granted, broadcast burning is still subject to review by the FFS and can be denied in the future.

TIPS ON CONTROLLED FIRE USE IN MEADOWS

If you or your contractor meets the legal requirements to conduct a prescribed burn, these are some of the things you should consider when burning a meadow in Florida.

- **Proximity to other potentially flammable objects** and plants outside the meadow: During dry periods embers can spread to other locations.
- **Creation of a fire break** or method to control the spread of fire: Typically, fire breaks consist of tilled lines, but sometimes they may be mowed or scalped buffers that are wetted down with water prior to ignition.
- **Height of the meadow vegetation**: This can dramatically alter the intensity and spread of a fire. You may choose to mow the meadow prior to ignition to reduce flame intensity and increase control.
- **Fuel type**: A meadow that consists primarily of grasses will generally burn hotter and faster than a meadow that contains more forbs. If the meadow features a pine overstory, this will increase the flammability of the site, whereas non-pyrophytic oaks (such as laurel oak, live oak, water oak) in the area will reduce flammability.
- **Wind direction**: Be mindful of wind direction, as this can significantly affect fire behavior by creating a **heading fire** (a fire moving with the wind) versus a **backing fire** (fire going against the wind). A heading fire will be far more intense and less controllable.
- **Fire season**: The seasonal timing of fire can greatly influence the resulting composition of a meadow, especially with regard to the flowering of fire-adapted forbs and grasses. See the next section for more information on specific effects and considerations for fire season.

Controlled burning of a meadow with many trees. Photo by Gage LaPierre.

Winter	Spring	Summer	Fall
December-February	March-April	May-August	September - November
✓ Weather is frequently conducive for burns ⚠ Does not stimulate summer or fall flowering species ⚠ May harm nesting wildlife populations ⚠ May harm annual seedlings that germinate during this period	✓ Will stimulate fall flowering species ✓ Knocks back woody and weedy fibrous species ⚠ May harm annual wildflower species that are flowering	✓ Will stimulate fall flowering species ✓ Will reduce meadow structure height going into fall & winter ✓ Knocks back woody and fibrous species ⚠ Frequently too wet for clean burns ⚠ May harm younger forbs (one to two years of age)	⚠ Will harm nesting wildlife populations ⚠ Will harm fall flowering species

Pros and cons of prescribed burning by season. Illustration by Isabella Guttuso Browne.

BURNING SEASONALITY CONSIDERATIONS

Late spring to midsummer are ideal times for burning meadows in Florida. This period avoids harming spring wildflower stands and encourages the flowering of fall species. Winter burning does little to encourage the flowering of fall-flowering species, and it can harm the seedlings of spring annuals. However, the winter season often offers more days with favorable weather conditions for burning and produces similar favorable outcomes compared to spring or summer burning. Generally late-summer burning is not harmful, but the weather is often too wet to perform clean burns. Burning during fall months is not advised because it harms fall-flowering species as well as insects and wildlife that are taking up nesting for winter.

Herbicide

Herbicide is an excellent tool in managing meadows, but like any tool it can easily be misused. The primary reason to use herbicides in managing meadows, outside of site preparation, is to control invasive or weedy species. Rarely should management treatments ever consist of more than spot applications. Ideally, individuals tasked with managing wildflower meadows should be certified to spray herbicide consistent with state law and product labeling; keep in mind that it is illegal to use herbicide outside of label limits. Herbicide is best applied when weeds are actively growing. In Florida, this generally means herbicide applications are not as effective during plant dormancy (winter for warm-season weeds) or during drought conditions. Midday applications are generally most effective; however, care should be taken to meet the label's recommended rain-fast period; check weather forecasts to avoid afternoon rain showers that can wash the herbicide off the cuticle of the plant.

Mulch

Various types of mulch, including bark, cypress, rubber, pine straw, and wood mulch are commonly used in landscaping, but their application in meadow establishment and maintenance requires careful consideration. In traditional landscaping, wood mulch is often employed for weed suppression, moisture retention, soil temperature regulation, and aesthetic appeal. However, these benefits don't always translate positively to meadow ecosystems. While wood mulch may initially suppress weeds, it may also inhibit the germination and growth of desirable meadow species, which often thrive in sandy, nutrient-poor environments. Furthermore, while mulch does retain moisture, it can create overly moist conditions or conditions that favor weeds and fungal diseases, particularly in meadows designed for well-drained soils. Soil temperature regulation is also more critical in formal landscaping than in meadows, where a diverse plant canopy naturally moderates temperature fluctuations. Mulch also adds organic matter, potentially increasing soil fertility, which can promote the growth of weedy species. Finally, while wood mulch can create a neat, uniform appearance, this aesthetic contrasts sharply with the natural, dynamic look of a meadow. Therefore, the traditional benefits of wood mulch are often counterproductive in meadow settings, where the goal is to foster a self-sustaining, diverse ecosystem that mimics natural conditions.

Unlike other types of mulches, pine straw is one exception that can be a valuable tool during the initial establishment phase of a meadow. Its light, airy texture allows for good seed-to-soil contact and can help retain moisture, aiding in germination. Pine straw can also serve as an effective fuel source for prescribed burns, facilitating the spread and even coverage of fire. Once the meadow is established, pine straw mulch should no longer be necessary, though it can be applied effectively to carry a prescribed fire through a meadow. For long-term meadow health, a light layer of straw or no mulch at all allows for better seed-to-soil contact, promotes natural soil conditions, and does not hinder the growth of meadow species.

Meadow Dormancy

Meadows will experience a period of dormancy in the winter months of Florida (November to February). Most plant species will slow down or stop their growth during this period. Plant dormancy is a survival mechanism triggered by environmental cues like decreasing temperatures and shorter daylight hours. These signals initiate a series of physiological changes within the plant, including the cessation of growth, shedding of leaves, and reduction in metabolic activity. This state of dormancy allows plants to conserve energy and resources during potentially harsher winter conditions, protecting them from freezing temperatures, water scarcity, and other stressors. The length and extent of dormancy will vary depending on the location in the state and plant species; some species in South Florida may experience limited dormancy or none at all.

Ecological Benefits of Dormancy

Dormant meadows, often overlooked during winter, are teeming with life. These seemingly lifeless landscapes provide essential shelter and food for a

left A bee burrows into the hollow center of a plant for winter dormancy. Photo by Hectonichus, 2009. Creative Commons license via Wikimedia Commons, https://commons.wikimedia.org/wiki/File:Apidae_-_Ceratina_cyanea-001.jpg.

right Insect during dormancy. Photo by zygy, 2021. Creative Commons license via iNaturalist, https://www.inaturalist.org/observations/68192382.

diverse array of wildlife, from birds and small mammals to beneficial insects. The stems of flowers and grasses, for example, provide nesting habitat for numerous species of bees, moths, and butterflies that, in turn, help feed birds and other wildlife.

Aesthetics of Dormancy

Some people consider meadows less attractive during the winter months due to the browning of plant material and its dead-like appearance. Many other people, however, view meadows in wintertime as offering a unique aesthetic appeal that highlights the cyclical nature of life. How one sees aesthetics like these is a matter of opinion and perspective that is shaped by life experiences, social influences, and knowledge. It is often difficult to change social or individual aesthetics without education and long-term exposure. This is a consideration for homeowners deciding whether to install a meadow.

Dormancy Management Actions

Management during the dormant period might be neccesary to achieve a specific objective or goal, such as counteracting undesirable weeds. However, management activities (including mowing or burning) during the dormant season are largely discouraged because they can harm hibernating or developing insects. If management actions are taken in the fall, leave part of the meadow undisturbed and 6-inch stubble in the cut areas to retain some benefits for fall-nesting pollinators and other insect populations.

Quality

When managing a meadow over the long term, the question of what constitutes a "good-quality" meadow inevitably arises. The concept of quality, like many things, is semi-subjective and depends largely on the observer's goals and chosen metrics. At its core, a meadow's quality can be evaluated by how well it meets the objectives set for it. However, comparing meadows to one another presents a challenge.

In ecological restoration, quality is often assessed by examining the community composition and its resemblance to a reference site. However, as inherently variable and typically human-made ecosystems, meadows often

A residential meadow during dormancy. The designed nature of the meadow is apparent even in dormancy through the varying heights, shapes, and shades of brown and gray. Photo by Benjamin Vogt.

A meadow during dormancy. Photo by Gage LaPierre.

lack natural reference sites for comparison. This variability necessitates an alternative approach to quantifying their quality.

One widely used system in restoration ecology is conservation quality assessments, which rely on a metric called **coefficients of conservatism** (C-**value**). These coefficients assign numerical values to plant species based on two criteria: (1) their fidelity to specific habitats, and (2) their tolerance for disturbances. Ruderal or weedy species, those commonly associated with human disturbances, are given low values, whereas species found in undisturbed or sensitive habitats receive high values. By leveraging this system, we can quantitatively evaluate the compositional quality of a Florida meadow. C-values for Florida can be found through online resources at The Institute for Regional Conservation website.[4] While this database focuses on South Florida species, there is a great deal of overlap with other regions in the state. Efforts are underway to update coefficients of conservatism by region across the state of Florida.

Steps for Assessing Meadow Quality

1. **Conduct a species inventory**: Begin by creating a comprehensive list of all plant species present in the meadow.
2. **Calculate the mean coefficient of conservatism (C-value)**: Using pre-assigned coefficients for each species, determine the mean C-value by summing the coefficients of all species and dividing by the total number of species.
3. **Compute the Floristic Quality Index (FQI)**: To calculate the FQI, multiply the mean C-value by the square root of the total number of native species in the meadow:

$$FQI = C \times \sqrt{n}$$

This index provides a standardized, quantitative measure of meadow quality that allows for objective comparisons across sites. Following this approach allows the quality of meadows to be evaluated in a consistent and nonsubjective manner, aiding in their restoration and management. The main flaw of this method is that it does not account for the varying impacts of nonnative species, treating them all as detrimental regardless of their ecological integration. This method is also biased against ruderal or weedy species in favor of select native species richness, which undervalues the ecological roles and functionality of ruderal species. Still, this is one of the few, perhaps the only, quantitative method for assessing the quality of meadows quickly.

For small-scale or residential meadows, the necessity for rigorous quality measurement depends heavily on the client's or homeowner's goals. While a standardized index can provide valuable data, it may not always align with the priorities of a backyard meadow enthusiast. Residential meadows often prioritize aesthetic appeal, personal enjoyment, and the creation of habitat for specific wildlife, such as pollinators. In these cases, qualitative observations—such as the presence of desired plant species, the abundance of pollinators, or the overall visual appeal—may be more relevant than strict quantitative metrics. Homeowners might focus on documenting the return of specific bird species or simply observing the progression of flowering plants throughout the seasons. Therefore, while quantitative assessments offer valuable insights for large-scale meadow projects, a more flexible and personalized approach to evaluating success may be better for smaller meadows.

five

CASE STUDIES

This chapter provides six case studies of meadows installed in Florida. The planning, design, installation, and maintenance are described with the best detail available.

Lake Nona Nemours Children Hospital Meadow

This roughly 30-acre meadow is located in Orlando, Orange County, Florida.

Goals

The three main goals for the project were:

- to reduce the irrigation and maintenance demands of landscaping;
- to provide aesthetically pleasing views for patients; and
- to restore a sense of place by mimicking a mesic flatwoods community—the dominant plant community in Central Florida.

Objectives

- Limit the reestablishment of invasive species to less than 5 percent coverage.
- Achieve a diverse stand of native forbs and grasses.

Soils

Spodosol—Hyperthermic Aeric Alaquods—Smyrna soil series. During the construction of the hospital, the natural soil was overburdened with waste

opposite Prairie coneflower (*Ratibida pinnata*). Photo by Lilly B. Anderson-Messec.

Meadow with the Nemours Children's Hospital in the background.
Photo by Nancy Bissett, The Natives.

soil by 3–4 inches. This soil was very sandy in some areas and very clay dominated in others. The pH across the site was also high.

Site History

In its natural state, this site was a mesic flatwoods community, which was converted first into a citrus grove and then into an improved pasture dominated by bermudagrass, bahiagrass, and cogongrass. Prior to the construction of the Lake Nona urban complex, much of the land was graded with backfill (1–4 ft.) from the construction of several large retention basins in the area. Construction of Nemours Children's Hospital began in 2009, and the hospital opened in 2012. This meadow project started in 2010, two years prior to the opening.

Planning and Design

Prior to the onset of planning for this project, the clients were shown an overview of an installation and maintenance plan. The manager of this project perceived this step to be important because the clients' perceptions of the timeline and desired aesthetic outcomes did not match the project proposal. Meetings were also held with the construction companies that were operating within the area to ensure that there would be limited conflict with the development of the meadow project due to debris storage, equipment movement, and the like. This proved to be an unexpectedly valuable step because a construction company agreed to help provide temporary irrigation for the site. The project was also broken into two units: one in front of the hospital and the other behind it.

One of the primary design requirements in this project was to ensure the meadow in the front of the building featured showy blooms. A secondary requirement was to ensure there was little bare ground in areas close to the sidewalks and the road to limit erosion concerns from runoff. To meet these requirements, the design of the meadow included placement of additional containerized plant material and seed along these key areas.

Site Preparation

The site had been overladen with a deep mix of fill material. In some areas, the soil was heavy in clay and in others extremely sandy. To alleviate this variability in the soil profile, fill material was brought in and spread across

Site preparation for the front unit of the Nemours Children's Hospital meadow, showcasing (*left*) the area where clay was thick, and (*right*) the soil mixture used in the front unit. Photos by Nancy Bissett, The Natives.

The front unit of the Nemours Children's Hospital meadow (*top*) directly following planting, and (*bottom*) six months following planting. Photo by Nancy Bissett, The Natives.

both units. In the back unit, fill sand was spread across the site at thicknesses from 2 to 4 inches. For the front unit, where a denser stand of wildflowers was desired, a soil mixture was developed to aid germination. This mixture was laid down in layers, with the bottom layer consisting of 4 inches of clean sand, the middle layer of 1 inch of Florida peat, and the top layer of 1 inch of composted material. Next, the top 3 inches of soil was lightly disced to mix the layers. The purpose of including the compost material was to help prevent erosion and retain moisture for seedling germination.

Site Installation

Installation took place in the winter season of 2010. More than 78 different indigenous species, including 31 native graminoid species, were seeded and planted over 30 acres. In total, more than 400 pounds of seed was used across the project site, roughly 50 seeds per square foot. Containerized planting materials consisted of 36,000 one-gallon plants, roughly one plant for every 36 square feet. Roughly 60 percent of the plants were placed in the front unit and along key sidewalk areas. Seeds were installed using a Grasslander tractor-drawn seeder. The back section was installed first, and the front section followed a year later. Directly after installation, both units were provided with temporary irrigation that aided in germination.

Maintenance

Following installation, trained crews went out with backpack sprayers to target problematic weeds, including bermudagrass (*Cynodon dactylon*) and cogongrass (*Imperata cylindrica*). Both units of the meadow were mowed only once in the first year, during midsummer, using a tractor-drawn rotary

The back unit of the Nemours Children's Hospital meadow (*left*) several weeks following planting, and (*right*) six months following planting. Photos by Nancy Bissett, The Natives.

Species in the 400-Pound Seed Mixture

Agertina jucunda
Andropogon brachystachyus
Andropogon glomeratus
Andropogon ternarius
Andropogon decipiens
Andropogon glauca
Aristida spiciformis
Aristida stricta
Aster adnatus
Aster dumosus
Aster subulatus
Aster tortifolius
Aureloaria pedicillata
Buchnera americana
Carphephorus carnosus
Carphephorus corymbosus
Carphephorus paniculatus
Carphephorus subtropicanus
Chrysopsis mariana
Coreopsis floridana
Coreopsis leavenworthii
Ctenium aromaticum
Digitaria filiformis
Elephantopus elatus
Eragrostis elliottii
Eragrostis spectabilis
Eragrostis virginicus
Eriocaulon decangulare
Eryngium yuccifolium
Eupatorium mohrii
Eupatorium rotundifolium
Euthamia caroliniana
Galactia volubilis
Gymnopogon chapmanianus
Helianthus angustifolius
Helianthus carnosus
Helianthus radula
Hieracium megacephalon
Hypericum reductum
Hyptis alata
Liatris gracilis
Liatris laevigata
Liatris spicata
Lobelia glandulosa
Ludwigia maritima
Muhlenbergia capillaris
Oxypolis filiformis
Panicum anceps
Panicum hians
Panicum longifolium
Panicum rigidulum
Panicum verrucosum
Pityopsis tracyi
Polygala rugellii
Pterocaulon pycnostachyum
Pterocaulon virgatum
Rhexia cubensis
Rhexia mariana
Rhexia nashii
Rhexia petiolata
Rhynchospora colorata
Rhynchospora fascicularis
Rhynchospora inundata
Rhynchospora latifolia
Rhynchospora microcarpa
Rhynchospora plumosa
Rudbeckia hirta
Saccharum giganteum
Schizachyrium stoloniferum
Scleria reticularis
Solidago chapmanii
Solidago fistulosa
Solidago stricta
Sporobolus curtissii
Xyris ambigua

left The site three years following installation. Photo by Nancy Bissett, The Natives.

right The back unit three years following installation. Photo by Nancy Bissett, The Natives.

mower at a 6-inch mow height. In the following years, the meadow was mowed with the same mower but at a height of 10 inches and during the late-spring period. Starting around 2016, nearly six years after the site was installed, the meadow maintenance lagged and the meadow composition suffered due to a change in the contract for site management. One issue that arose in the northern part of the back unit was the introduction of invasive species such as cogongrass (*Imperata cylindrica*) and natal grass (*Melinis repens*), presumably via the decks of mowers operating along adjacent sod fields. The original management team resumed maintenance as of 2023 and began controlling these species.

Lessons Learned

- This case study demonstrated that large-scale meadows can be maintained effectively over a long period.
- However, a lag in maintenance and a deviation from the maintenance plan did create issues with an influx of invasive plants over a six-year period. Specifically, residual cogongrass appeared along the margins, and other invasives like rose natal grass were brought in by uncleaned mowing equipment and ill-timed mowing.

left The site in the winter of 2023, thirteen years after construction. Photo by Gage LaPierre.

right The site in the winter of 2023, thirteen years after construction. Photo by Gage LaPierre.

- As maintenance personnel and decision makers responsible for the property changed, reeducation in the purpose, goals, expectations, and maintenance of the meadow was needed to continue adherence to the long-term plan.

University of Florida Natural Area Teaching Laboratory

This one-acre, grant-funded project is located in Gainesville, Alachua County, Florida.

Goals

The two main goals for the project were:

- to enhance the site with a diversity of native species to improve wildlife/pollinator habitat for educational purposes; and
- to prevent invasive plant species recolonizing the site as much as possible.

Views of the NATL meadow, featuring many bunching grasses. Photo by Gage LaPierre.

Highlighting differences in soil across the NATL meadow site, (*left*) silt/clay-dominated soil in the western half of the field, and (*right*) sand-dominated soil on the eastern half. Photos by Gage LaPierre.

Objectives

Success was defined as 80 percent dominance of the site by desirable natives, including showy blooms, less than 5 percent coverage by invasive species, and less than 25 percent coverage by weedy species.

Soils

The soil type for this area was historically classified as Zolfo series, characterized by deep sands with a spodic horizon (a recognizable layer in the soil where organic matter and aluminum have gathered). However, the land-use history of this site featured extensive soil disturbance via refuse dumping and construction use, which created an excess of silt and clay across half of the site.

Site History

The Natural Area Teaching Laboratory (NATL) is a unique, 70-acre conservation and teaching area located on the University of Florida's main campus in Gainesville. The site proposed for this grant was a 1.05-acre block within a management unit known as the old-field successional area. Prior to NATL

In 2018 the eastern side of the NATL field was dominated by cogongrass (*Imperata cylindrica*), and Guinea grass (*Megathyrsus maximus*). Photo by Gage LaPierre

In 2018 the western side of the NATL field was dominated by ragweed (*Ambrosia artemisiifolia*) and Spanish needles (*Bidens alba*). Photo by Gage LaPierre.

being actively managed, the old-field area was used first for cattle grazing, then as a nursery refuse area, and then finally as a construction dump site. As a result, the flora were dominated by several different invasive species, including cogongrass (*Imperata cylindrica*), Guinea grass (*Megathyrsus maximus*), and Johnsongrass (*Sorghum halepense*), as well as weedy plants such as hairy indigo (*Indigofera hirsuta*) and bermudagrass (*Cynodon dactylon*). NATL management successfully eradicated these species via herbicide in 2019, two years prior to starting the project in 2021. During the 2020 growing season, the plant composition of the site was dominated by weedy ruderal species such as ragweed (*Ambrosia artemisiifolia*), Spanish needles (*Bidens alba*), wild lettuce (*Lactuca canadensis*), showy rattlebox (*Crotalaria lanceolata*), black medic (*Medicago lupulina*), and wild radish (*Raphanus raphanistrum*). In the winter season of 2021, the site was mowed using a tractor-drawn rotary mower, after which nonnative Italian ryegrass (*Lolium multiflorum*) dominated the site. In the spring of 2021, directly prior to the project starting, large patches of bermudagrass (*C. dactylon*) were present across the site.

Planning and Design

During the planning phase, several experts in the region were consulted on site preparation and maintenance methods. Soil samples were taken across the site to map where thicker clay deposits were located. This information was used to guide the installation design. A native landscape design company in Gainesville donated their time to create a planting design.

The primary design factor in this meadow was to craft a planting scheme based on the soil mapping done earlier. Another factor was to ensure that the front slope featured a high concentration of tickseed (*Coreopsis* spp.) for showy spring blooms visible from the road.

Site Preparation

For this project the site was prepared using the repetitive cultivation technique. This decision was made to meet grant requirements that restricted the use of herbicide in site preparation. Preparation began in early May 2021 with mowing using a tractor-drawn rotary mower to knock down vegetation. Next, large debris, chunks of construction debris, and rocks were removed using a tractor-drawn stone/landscape rake, which produced a clean enough site to use a tiller. Next, a deep tillage of 8 inches was performed in late May via a tractor-drawn disc. This helped break up the ground and kill woodier perennial weeds. In June, shallow cultivation (1–2 in. depth) was then employed using a tractor-drawn tiller. Shallow cultivation continued throughout the summer and into the fall prior to plant installation. The frequency of cultivation was determined based on rain events and weed-seed emergence. The average cultivation frequency ended up amounting to every two weeks, with a total of ten cultivations.

The NATL field was mowed in winter of 2021. Photo by Gage LaPierre.

The NATL field in the summer of 2021: (*left*) The field being disced and (*right*) in the aftermath of discing. Photos by Gage LaPierre.

left In the winter of 2021, the field post-mowing was dominated by Italian ryegrass (*Lolium multiflorum*). Photo by Gage LaPierre.

right The site being prepared for planting in November 2021. Photo by Gage LaPierre.

Site Installation

The site was seeded and planted in November 2021 with the help of many volunteers. The 12-pound seed mixture consisted of a seed mixture harvested from Austin Cary Forest and purchased from the Florida Wildflowers Growers Cooperative. Seeds were mixed with building sand to reduce clumping prior to broadcasting (five parts sand to one part seed). Seeds were subsequently rolled into the soil using a 310-pound sod roller. More than 10,000 containerized plants were then installed across the site. These consisted primarily of 10-inch native bunchgrass plugs, along with several hundred one-gallon potted forbs. Planting design was determined primarily by soils and shading. Prior to and following the planting date, there was a significant rainfall event, which presumably helped in plant establishment. Post rainfall, however, Italian ryegrass (*Lolium multiflorum*) germinated far more densely than anticipated and likely outcompeted some seeded species.

Maintenance

Following installation, Italian ryegrass, which is a winter annual species, germinated very heavily and dominated much of the site. By March 2022 it was approaching a 2-foot average growth height. In an effort to reduce

top The site after the ground has been prepared in November 2021. Photo by Gage LaPierre.

bottom left Containerized plants staged on the site during planting in November 2021. Photo by Gage LaPierre.

bottom right The site during planting in November 2021. Photo by Gage LaPierre.

left The site in December 2021 directly following planting. There is dense germination of Italian ryegrass (*Lolium multiflorum*). Photo by Gage LaPierre

right The site in December 2021 showing dense new growth of Italian ryegrass. Photo by Gage LaPierre.

the dominance of this species and to prevent it from reseeding and causing future issues on the site, we implemented a mowing treatment using a tractor-drawn flail mower at a height of 6 inches. The timing of this mowing treatment proved effective at reducing and preventing the reseeding of Italian ryegrass, which as an annual cool-season species, dies off when warmer temperatures arrive.

In the summer of 2022, different species of native bunchgrasses and sedges quickly began to fill in across the site. Native species of tickseed (*Coreopsis*) flowered in patches, along with other forbs. Many forbs, including black-eyed Susan (*Rudbeckia hirta*), cutleaf coneflower (*Rudbeckia laciniata*), purple coneflower (*Echinacea purpurea*), and silkgrass (Pityopsis spp.), did not flower but grew in across the site. In the late summer, **ruderals** (pioneer species), including ragweed, began to dominate the site. To reduce this species' dominance and prevent reseeding, we mowed the site down to 8 inches using a self-powered sickle bar mower.

In the fall of 2022 several species of native bunchgrasses flowered and

Germination of seeded species on the wetter side of the field. Photo by Gage LaPierre.

The site in May 2022, two months following mowing. The brown material is dead Italian ryegrass. Photo by Gage LaPierre.

right The site in July 2022, showing tickseed (*Coreopsis leavenworthii*) in bloom. Photo by Gage LaPierre.

below left Passionflower (*Passiflora*) and tickseed (*Coreopsis*) in bloom. Photo by Gage LaPierre.

below right The site in July 2022, showing black-eyed Susan (*Rudbeckia hirta*) not yet flowering but established from seed, along with native sedges (*Carex* spp.). Photo by Gage LaPierre.

left The site in August 2022, featuring high growth of ragweed (*Ambrosia artemisiifolia*). Photo by Gage LaPierre.

right The site in August 2022, after it had been mowed using sickle bar mower, with the shorn vegetation left on-site. Photo by Gage LaPierre.

seeded across the site. Partridge pea (*Chamaecrista fasciculata*) flowered, but blooms of blazing star (*Liatris*) and silkgrass (*Pityopsis*) were sparse. Spanish needles (*Bidens alba*) and ragweed (*Ambrosia artemisiifolia*) became dominant in certain areas of the site. To reduce the coverage of these species, we employed a STIHL reticulating sickle bar articulating hedge trimmer to spot-treat stands, cutting them down to a height of 6 inches. The site was mowed once in the late summer of 2023 per the management plan. As of the fall of 2023, ocular surveys revealed the site contained roughly 90 percent desirable native species dominance with less than 1 percent coverage of invasive species and less than 10 percent nonnative species.

Lessons Learned

- The strong dominance of ryegrass early on likely smothered many of the seeded forbs and potentially contributed to the influx of weedy species in year two due to low establishment of dense stands of showier species like tickseed (*Coreopsis* spp.) and black-eyed Susan (*Rudbeckia hirta*).

aboveright and left The site in October 2022, after the mowing treatment in August 2022. Photo by Gage LaPierre.

right The site in spring 2023: Lanceleaf tickseed (*Coreopsis lanceolata*) is in bloom. Photo by Gage LaPierre.

left The site in spring 2023 displays mixed native graminoids. Photo by Gage LaPierre.
right The site in spring 2023 sports wildflowers in bloom. Photo by Gage LaPierre.

- Without the installment of graminoid plugs, this meadow project would have a been a failure.
- The repeated-cultivation site preparation method did work well to flush out warm-season weeds from the seed bank. Prolonging this treatment into the winter season (late November or December) might have reduced winter weed-seed emergence.
- In one low-lying area, repeated cultivation failed to control bermudagrass effectively; if it continues to spread, this species may present problems for the meadow composition in years to come.

Species seeded and containerized plants installed at the Natural Area Teaching Laboratory

Scientific Name	*Common Name*
SEEDED SPECIES	
Coreopsis lanceolata	Lanceleaf tickseed
Coreopsis leavenworthii	Leavenworth's tickseed
Eragrostis elliottii	Elliott's lovegrass
Heliopsis helianthoides	False sunflower
Asclepias tuberosa	Butterfly milkweed
Echinacea purpurea	Purple coneflower
Chamaecrista fasciculata	Partridge pea
Desmodium perplexum	Perplexed ticktrefoil
Liatris squarrosa	Scaly blazing star
Tephrosia virginiana	Goat's rue
Eryngium yuccifolium	Rattlesnake master
Sorghastrum nutans	Indiangrass
Schizachyrium scoparium	Little bluestem
Panicum virgatum	Switchgrass
Verbesina virginica	Frostweed
Vernonia gigantea	Ironweed
Ratibida pinnata	Gray headed coneflower
Solidago altissima	Tall goldenrod
Symphyotrichum leave	Smooth aster
Sporobolus compositus	Tall dropseed
Rudbeckia hirta	Black-eyed Susan
Achillea millefolium	Yarrow
Eutrochium fistulosum	Joe-pye weed
Liatris spicata	Spiked blazing star
Elymus virginicus	Virginia wildrye
Chasmanthium latifolium	River oats
Asclepias incarnata	Swamp milkweed
Bidens frondose	Tickseed sunflower
Andropogon gerardii	Big bluestem
Tradescantia ohiensis	Spiderwort

Dichanthelium clandestinum	Deertongue grass
Rudbeckia triloba	Brown-eyed Susan
Conoclinium coelestinum	Mistflower
Penstemon digitalis	Smooth beardtongue
Helenium autumnale	Sneezeweed
Pycnanthemum incanum	Hoary mountain mint
Lobelia cardinalis	Cardinal flower
Mimosa quadrivalvis	Sensitive mimosa

CONTAINERIZED PLANTS

Aristida beyrichiana	wiregrass
Asimina obovate	Common pawpaw
Baptisia alba	White indigo
Eragrostis elliottii	Elliott's lovegrass
Penstemon multiflorus	Many-flower penstemon
Prunus umbellate	Flatwoods plum
Salvia coccinea	Blood sage
Sorghastrum nutans	Indiangrass
Solidago odora	Sweet goldenrod
Tridens flavus	Purple top grass
Eryngium integrifolium	Snakeroot
Helianthus angustifolius	Swamp sunflower
Hypericum brachyphyllum	St. John's wort
Ilex glabra	Gallberry
Liatris spicata	Marsh blazing star
Lyonia lucida	Fetterbush
Muhlenbergia capillaris	Muhly grass
Sisyrinchium spp.	Blue-eyed grass
Tripsacum dactyloides	Gamagrass
Cephalanthus occidentalis	Button bush
Hibiscus coccineus	Scarlet hibiscus

Troy Springer Residential Meadow. Photo by Troy Springer.

Troy Springer Residential Meadow

This 1.3-acre meadow is located in Plant City, Hillsborough County, Florida.

Goals

The four main goals for the project were:

- to provide a demonstration garden model for business marketing and clientele development;
- to support a wide range of wildlife;
- to provide a seed source for building other meadows for clients; and
- to form part of a broader beautification project for Springer's personal residence.

Objectives

Success was defined as two-to-one native forb–to–grass coverage with 95 percent coverage of desirable species and less than 5 percent weed cover.

Soils

Entisol—Typic Quartzipsamments—Gainesville series. The soil was typical for the area, consisting of deep, well-draining yellow sands well suited for citrus and peanut crop production.

Site History

Based on old aerial photographs and the soils present, this site would likely have contained a sandhill plant community. In the early twentieth century, the area was converted to farming for citrus. Today, the site is nested in a semirural residential area. Prior to the start of this project, the site was dominated by sparse, dilapidated remnants of citrus trees and a lawn of mixed St. Augustine grass (*Stenotaphrum secundatum*), bahiagrass (*Paspalum notatum*),

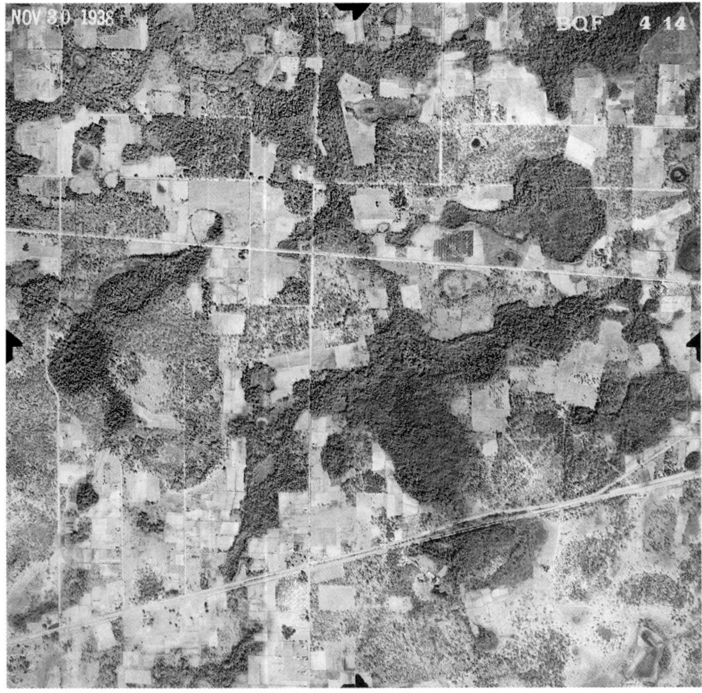

Historical imagery of the site in 1938. Photo by the US Department of Agriculture. Courtesy of the University of Florida, George A. Smathers Libraries Digital Collections.

and centipedegrass (*Eremochloa ophiuroides*). Parts of the site featured sparse ornamental landscaping and three large laurel oaks (*Quercus laurifolia*), which were all removed. A few other mixed hardwoods remained.

Planning and Design

The project site sits on two parcels of land. The western parcel (0.85 acres) was constructed first and consists of three sections. The eastern parcel (0.45 acres) was constructed roughly three years later and is essentially one large meadow. The eastern section, which is the front yard, receives full sun, while the western section, which are the side yards and backyard, is slightly more shaded. To keep this project manageable in terms of time and costs, both parcels were installed in phases that grew over several years. In other words, at the start of this project the scope for both sections was roughly a third of the total area that they take up today. Planning in phases allowed the meadow to grow outward naturally as more areas were prepared.

One of the main design features of this project included placement of view-enhancing shrubs as well as several longleaf pines (*Pinus palustris*) during each expansion period. Another design aspect was to incorporate curves throughout the meadow, along with a buffer between the meadow and the remaining turfgrass lawn.

Site Preparation and Installation

This project employed the sod-removal technique, which was performed in gradual phases over the course of seven years. The sod was removed in the spring, then weeds were controlled by hand. In the western section, the area was planted in the spring following site preparation with many containerized plants spaced on 3–4-foot centers to minimize cost. To expand diversity, additional seed from the Florida Wildflowers Growers Cooperative was mixed in with seed harvested on-site. Over the following years, the western meadow was expanded by a small amount annually until it reached today's footprint, using the same methods as in previous years.

The eastern meadow was constructed using similar methods as for the western site. The main difference was that a much larger area was initially planted compared to all previous plantings, so that the current footprint (two sections) took only two years to complete. Another difference was that, while

The Springer meadow one year post-installation. This project was gradually expanded over time. Photo by Troy Springer.

The western side of the meadow one year post-installation. It, too, was gradually expanded over time. Photo by Troy Springer.

some 4-inch plant material was used, the eastern meadow relied more heavily on seeding than containerized materials. Roughly 6–8 pounds of seed per acre was sown, and a light layer of pine straw was spread immediately after sowing. The seed mixture used consisted of spiked blazing star (*Liatris spicata*), Maryland goldenaster (*Chrysopsis mariana*), lanceleaf tickseed (*Coreopsis lanceolata*), lovegrass (*Eragrostis* spp.), narrow-leaf silkgrass (*Pityopsis graminifolia*), blanket flower (*Gaillardia pulchella*), and black-eyed Susan (*Rudbeckia hirta*). After the seed was sown, the ground was covered with a light layer of pine straw. In both sections, a small quantity of assorted containerized plants was also installed. All meadow expansions included installation of additional containerized wildflowers and grasses to boost diversity and experiment with what species would grow best on-site.

Site Maintenance

Post-installation neither the western nor eastern section was irrigated, except for watering the containerized plants until established. In the following year, both sections were mowed in late winter using a zero-turn rotary mower down to 2-inch stubble (excluding shrubs). Several passes were made with the mower to chop up the very small clippings, and most of the clippings were left on-site except where the debris appeared thick enough to choke out desired plants. In the second year an additional mowing treatment was performed. Across the first two years, intense hand-weeding of nonnative grasses, sedges, and weedy forbs like Spanish needles (*Bidens alba*) was a focus. By the third year, much less weeding was needed to keep all the meadows clean and meet the high aesthetic goals established.

In the fifth year, portions of both meadows were burned during the late-winter season (February). In later years, burning occurred in late March and early April. Future burns will occur in some cases as late as June. Initially, burning occurred on one- to three-year cycles. However, experience has shown that alternating between a one-year and two-year cycle, as well as burning later in the season, results in less hand-weeding and a higher aesthetic quality. Currently, if a section of meadow is not burned, it will be mowed sometime between the end of March and the end of June, so that the entire site is either burned or mowed every year.

There have been occasions when aggressive, difficult-to-control nonnative grasses have entered the meadow. In those cases, herbicide is used to

The meadow during the first two years following installation. Photo by Troy Springer.

The site during the first two years following installation. Note the edging treatment on the lawn Photo by Troy Springer.

The meadow during a prescribed burn in 2023. Photo by Troy Springer.

The meadow regrowth following the prescribed burn in 2023. Photo by Troy Springer.

The meadow in 2024. Photo by Heather Crane, Native Plant Horticulture Foundation.

control them. Sparse cogongrass infestation is treated by swabbing each blade with glyphosate using a sponge in a gloved hand. Heavier infestations are sprayed directly with a backpack sprayer set on low pressure. In areas where infestation and removal of undesirable species results in widespread open areas, replanting occurs in the fall by adding either containerized plants or touch-up seeding. The edges of the meadow and largish open spaces inside the meadow are occasionally lightly mulched with pine straw to maximize appearance. Every time the turf adjacent to a meadow is mowed, the meadow is treated like any other landscape bed and is maintained with a power-driven edger to prevent turf from invading the meadow.

Lessons Learned

- Expanding a meadow in phases appears to be a very effective and less risky approach compared to installation of the entire meadow at once.
- Always clean mowing equipment thoroughly, as it appears likely that in this case mowing treatments introduced cogongrass and other weeds.
- Burning in the late-spring or early-summer months appears more effective than winter burning at reducing weed pressure.
- The sod-removal site preparation method requires either intensive follow-up weeding or a combination of additional site preparation approaches.

- Highly aesthetic meadows with low weed presence can be achieved even in formerly turf-covered areas.

Julie's Residential Meadow

This small, 2,300-square-foot residential meadow is located in Gainesville, Alachua County, Florida.

Goals

The four main goals for the project were:

- to reduce water and fertilizer input;
- to support pollinators and birds;
- to ensure ex-situ conservation of rare and threatened plant species, and
- to provide a relaxing view.

Julie's residential meadow two years following installation. Photo by Gage LaPierre.

A 1937 aerial image with a red circle highlighting where the project site is located. Photo by the US Department of Agriculture. Courtesy of the University of Florida, George A. Smathers Libraries Digital Collections.

Objectives

Achieve one-to-one native forb-to-grass coverage with 75 percent coverage of installed species.

Soils

Ultisol—Grossarenic Paleudults—Millhopper series—Deep sands with underlying layer of clay.

Site History

Based on old aerial photographs and the soils present, this site would originally have likely consisted of an upland pine plant community. In the early twentieth century this area was used for timber forestry and grazing. The neighborhood was installed during the 1960s, but the lot still contained some

The meadow site prior to installation and removal of laurel oaks.
Photo by Gage LaPierre.

remnant upland pine species such as longleaf pine (*Pinus palustris*), slimleaf pawpaw (*Asimina angustifolia*), sparkleberry (*Vaccinium arboreum*), and deerberry (*Vaccinium stamineum*). We worked around these species and retained them within the meadow during site preparation. The location where the meadow was installed in the yard had been maintained as an ornamental flower garden covered with pine bark mulch and containing indica azalea (*Rhododendron indica*), greenbriar (*Smilax* spp.), European holly (*Ilex aquifolium*), heavenly bamboo (*Nandina domestica*), and camellia (*Camellia japonica*). The site was also dominated by laurel oaks (*Quercus hemisphaerica*), which were removed or topped prior to meadow installation. Directly adjacent to the meadow is an emerald zoysia lawn (*Zoysia japonica* × *Z. tenuifolia* hybrid).

Planning and Design

The planning for this meadow consisted simply of a rough sketch of plug and seeding mixture placement. The primary design element in this meadow was to create soft curves along the boundary between the meadow and zoysia lawn. These borders would be separated with metal edging to help prevent the zoysiagrass from spreading into the meadow over time. Another key design feature was to place shorter grass species in the front to limit blocking the view of the house from the road.

Site Preparation

In this case, site preparation consisted of first removing all existing ornamental shrubs via a rubber-tire Kubota loader/excavator. Care was taken to limit disturbance to desirable remnant vegetation. Once this was done in the winter of 2020, the site was cleared of pine bark debris using a pitchfork and rake. Next, the organic material (duff buildup) and the top inch of the soil were removed from the site using the Kubota tractor mounted with a box blade attachment. The objective was to effectively remove the weed seed bank and provide bare mineral soil for seeds to germinate into. A portion of the meadow was leveled further with fill dirt brought to the site. This proved to be problematic later, as the fill dirt turned out to contain weed seeds.

Site Installation

The site was installed in January 2021 with approximately 8.4 pounds of clean seed purchased through a local supplier. A few dozen various-sized containerized plants were also installed. These mostly consisted of rare or threatened species. Prior to seeding, the site had received rain. The ground

left The project site in January 2021 following the removal of laurel oaks and site preparation. Photo by Gage LaPierre.

right The project site in January 2021 following site preparation and seeding. Photo by Gage LaPierre.

The meadow in the spring of 2021. Photo by Gage LaPierre.

The meadow in bloom during the spring of 2021. Photo by Gage LaPierre.

The meadow in the summer of 2021. Photo by Gage LaPierre.

The meadow in the summer of 2021. Photo by Gage LaPierre.

was prepared for direct seeding by dragging a hard rake across the ground, which left soft lines for the seeds to fall into. The seed was mixed with dry masonry sand, broadcast by hand, then pressed into the ground using a 200-pound sod roller. Following seeding the plants were installed using a punch dibble for plugs and a drill excavator for containerized plants. The site was irrigated for one week for 20 minutes total per day.

Maintenance

In the first year, the meadow was monitored for problematic weeds. Wedelia (*Sphagneticola trilobata*), English holly (*Ilex aquifolium*), and zoysia (*Zoysia* spp.) were all spot treated using glyphosate. The meadow was mowed in the winter of 2022 using a STIHL articulating hedge trimmer. Except for thicker stem material, most debris was left on-site. Later that summer, the site was mowed again using the same equipment. Larger stem material was taken off the site, while lighter material was burned off using a propane torch. The same maintenance treatments were applied again in 2023.

The meadow in 2022, showing dense growth. Photo by Gage LaPierre.

top left Julie's meadow in bloom. Photo by Gage LaPierre.

top right A close-up of Julie's meadow in bloom. Photo by Gage LaPierre.

bottom Julie's meadow in 2023, showing how the meadow design contrasts with and complements other landscaping elements. Photo by Gage LaPierre.

Lessons Learned

- Carefully check fill dirt being used on a meadow to ensure it is free of weed seeds.
- A noticeable decline in forb diversity and abundance occurred in the second year. This may have been the result of the use of fire in the first and second years. Fire, especially during the spring or summer as opposed to the winter, may damage forbs and favor grasses during the early phases of meadow development.
- Despite using a border guard, zoysiagrass from the lawn crept in slowly over the years. Monthly manual edging or application of a chemical barrier is required to keep zoysia and other rhizomatous species from creeping into the meadow over time.

Seed mixture recipe followed for Julie's Meadow

Scientific Name / Weight (oz.)			
Andropogon gerardii	3	*Galactia volubilis*	1
Andropogon glaucus	3	*Gaillardia pulchella*	5
Andropogon ternarius	2	*Gymnopogon ambiguus*	1
Andropogon virginicus	3	*Hypericum gentianoides*	4
Aristida beyrichiana	6	*Lespedeza stuevei*	2
Carphephorus corymbosus	6	*Liatris spicata*	4
Chamaecrista fasciculata	4	*Liatris tenuifolia*	4
Chrysopsis mariana	5	*Panicum anceps*	4
Coreopsis basalis	8	*Penstemon multiflorus*	2
Coreopsis lanceloata	6	*Pityopsis tracyi*	7
Coreopsis leavenworthii	6	*Rudbeckia hirta*	7
Ctenium aromaticum	6	*Salvia coccinea*	2
Digitaria filiformis	2	*Schizachyrium stoloniferum*	4
Elephantopus elatus	4	*Solidago odora*	4
Eragrostis elliottii	2	*Sorghastrum nutans*	3
Eragrostis spectabilis	1	*Sorghastrum secundum*	7
Eryngium integrifolium	1	*Tradescantia ohiensis*	1
Eryngium yuccifolium	1	*Tridens flavus*	7

An urban meadow installed at Encore! Solar Park in Tampa by Springer Environmental. Photo by Troy Springer.

Encore! Solar Park

This 0.35-acre meadow is located in Tampa, Hillsborough County, Florida.

Goal

The goal was to create an aesthetically pleasing urban green space.

Objectives

Success was defined as one-to-one native forb–to–grass coverage with 100 percent coverage of installed species.

Soils

Unknown; the site contained excessively disturbed urban soil.

Site History

Old aerial photographs of this site are not available and soil maps identify the site as disturbed urban soils, so it is not clear what the site condition was before development. Prior to the meadow installation the site was redeveloped as a large underground cistern used to capture rainwater for watering the entire landscape on the surrounding development. During this process, the previous soil was removed and replaced with fill dirt and topsoil. The soil present on-site had a significant amount of ground asphalt and concrete aggregate mixed in.

top The Encore! Solar Park site in January 2021, prior to site preparation and seeding. Photo by Troy Springer.

bottom The project site prior to site preparation in January of 2021 features monoculture turfgrass areas and hardscaping. Photo by Troy Springer.

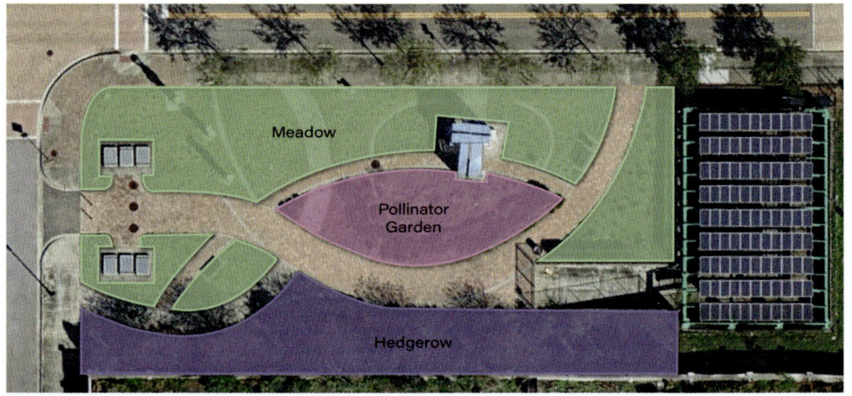

The overall design of the meadow, including a bisecting pathway, a central pollinator garden, and pavilion. Photo by Troy Springer.

Planning and Design

The planning for this meadow consisted of a detailed mapping of containerized plants drawn by a landscape architect. The builder of the project recommended making changes to the plant list because of the extraordinary site conditions, which consisted of a concrete floor just below the surface and the prevalence of ground asphalt and concrete in the soil. Some of those suggestions were not accepted by the property owner, so a compromise was reached. No seeds were used on this project due to a desire to greatly control initial plant placement.

The design of this project included a wide central walkway curving outward toward the southern border. A narrower pathway bisected the central area of the site, splitting it into a meadow along the northern border, a traditional landscape on the southern border, and a pollinator garden in the middle of the site.

Site Preparation

In this case site preparation consisted of removing all the old landscaping, which consisted of shrubs, pine bark mulch, and St. Augustine grass (*Stenotaphrum secundatum*). The removal was done using a small front-end loader and a team of workers. It was determined that the soil had too much organic matter, so the planting zone was amended with 12 yards of washed builder's sand.

Site Installation

Plantings for this project consisted entirely of containerized materials installed in February 2021. There were a total of 760 one-gallon plants consisting of the following mix:

bluestems (*Andropogon* spp.)
swamp milkweed (*Asclepius incarnata*)
small-flowered goldenrod (*Carphephorus corymbosus*)
lanceleaf tickseed (*Coreopsis lanceolata*)
purple lovegrass (*Eragrostis spectabilis*)
blanket flower (*Gaillardia pulchella*)
beach verbena (*Glandularia maritima*)
beach sunflower (*Helianthus debilis*)
spiked blazing star (*Liatris spicata*)
muhly grass (*Muhlenbergia capillaris*)
sea goldenrod (*Solidago sempervirens*)
sweet goldenrod (*Solidago odora*)
blue porterweed (*Stachytarpheta jamaicensis*)
Stokes aster (*Stokesia laevis*)
giant ironweed (*Vernonia gigantea*)

Maintenance

In the first year the meadow was monitored intensely and kept clean of all problematic weeds. The meadow was mowed in March 2022 using a zero-turn commercial mower at 3.75 inches high. Several passes were made with the mower to chop up the clippings to a small size. Prior to mowing, most tall wildflowers like blazing star (*Liatris*), giant ironweed (*Vernonia*), and goldenrod (*Solidago*) were cut manually with hand pruners to prevent the root balls from breaking loose from the soil as the mower passed over them. Prior to mowing, all seed remaining on plant stems was stripped or cut with pruners and redistributed into the meadow. All clippings from mowing were left onsite. A light layer of pine straw was added following mowing to help cover

top left The meadow in late winter 2022 during maintenance. Photo by Gage LaPierre.

top right The meadow in spring following mulching with pine straw. Photo by Gage LaPierre.

bottom The meadow in the summer of 2024. Photo by Gage LaPierre.

The meadow in the summer of 2024. Photo by Gage LaPierre.

up bare ground and alleviate the dead appearance following maintenance, which was not aesthetically appealing. In late February–early March of 2023 and 2024 the site was mowed again using the same tools and methods as in the first year. The design team and builder had intended to burn the meadow periodically, but the property owner has yet to allow this. The contractor who installed the meadow has continued monthly maintenance to keep the entire site looking its best.

Lessons Learned

- Expect higher project costs and more delays when working in dense urban areas.
- In a highly visited public setting like this, always set expectations with both the owner and also the visitors. Both must realize that there will be periods of dormancy and maintenance where the meadow

above The Tampa Encore! Solar Park meadow in the fall of 2024. Photo by Troy Springer.

right Another view of the Tampa Encore! Solar Park meadow in the fall of 2024. Photo by Troy Springer.

aesthetic is different from the spring or summer flowering periods. Signage helps.

- In dense urban areas, be aware that prescribed fire is difficult to plan for and the idea is frightening for owners who have little experience with a controlled burn.

The NaMa Native Landscapes Residential Meadow

This small, 800-square-foot meadow in a residential backyard is in Miami, Miami-Dade County, Florida.

Goal

To enhance the site with native species as a demonstration area for customers.

Objective

Success was defined as 90 percent dominance of the site by desirable natives (70 percent grasses and 30 percent forbs), including showy blooms, and less than 5 percent coverage of weedy species.

Soil

Unknown: excessively disturbed sandy/limestone urban soil.

Site History

Old aerial photographs of this site are not available. Soil maps show the site simply as disturbed urban soils, so it is not clear what flora was present on the site prior to development. Before the meadow was installed, the site contained a lawn and garden, consisting of a mix of nonnative turfgrasses and weeds.

Planning

Prior to the inception of this project in 2020, regional experts were consulted regarding species choices. The site consists of a roughly 800-square-foot area

A colorful meadow in an urban residential lot in Miami. Photo by NaMa Native Landscapes.

Another view of the meadow in an urban residential lot in Miami. Photo by NaMa Native Landscapes.

The NaMa site prior to meadow enhancement. Photo by NaMa Native Landscapes.

within a residential backyard. The longer, southern edge of the meadow is bordered by a fence line and hedges, whereas the shorter northern edge is bordered by a mixed polyculture lawn. Additional hedges and trees were planted along the fence line prior to meadow installment to increase greenery and privacy. Plants were to be sourced locally from what was available on the local native nursery market. Following the advice of experts in the region, such as Dr. Craig Hugel, the meadow was planned to start small and expand over time based on the owner's available time and desire. The first phase of the meadow consisted of a 25-foot-by-10-foot section, followed by a 10-foot-by-8-foot section in phase two. Both phases featured the same site preparation methods but differed in terms of species installed because of lessons learned from the first meadow.

Design

Key design elements in this meadow included soft curving borders, dominant use of different native grass species for ease of maintenance and to mimic local natural areas, and only locally available species. A layering effect was used to transition from the low-cut lawn to the hedges. Plant placement was based on a professional landscape design plan.

A close-up of the NaMa meadow three years following installation. Photo by NaMa Native Landscapes.

Site Preparation

Limited site preparation was performed and the anticipated need for intense weed removal was accepted due to the small size of the site. The site was prepared in the late winter of 2020 via the sod-removal method using a spade shovel rather than a sod-cutting device. Once the sod was removed, the site was left idle for two weeks to allow weed seeds to emerge and be removed. Following this treatment, plant materials were installed. This method was repeated for the second phase in the following year.

Site Installation

The site was planted in two phases with the first phase beginning in early spring 2021 using entirely containerized plants. No seed was used on this project in an effort to highly control initial plant placement and establishment. More than 125 plants were installed on roughly 18-inch centers, consisting primarily of plugs with a limited number of one-gallon potted plants. The plants were watered daily for the first two weeks to reduce mortality. Plants that died within the first six months were replaced as needed. One year later, in the second phase, roughly 45 plugs were installed.

Maintenance

The meadow was allowed to grow during the first year, then the following January, it was cut back by hand using a machete. Some species, such as scarlet sage (*Salvia coccinea*) and prairie clover (*Dalea feayi*) were removed because they had become too large and abundant, affecting the compositional and aesthetic qualities of the meadow. Seashells were added in some areas along the meadow border to provide a clean aesthetic and facilitate maintenance along the turfgrass border.

A wider view of the site three years following installation showing the contrast between meadow and lawn elements. Photo by NaMa Native Landscapes.

Containerized plants installed in the NaMa residential meadow by phase

First Phase	*Second Phase*
COMMON NAME (SCIENTIFIC NAME)	
Wiregrass (*Aristida beyrichiana*)	Wiregrass (*Aristida beyrichiana*)
Lanceleaf tickseed (*Coreopsis lanceolata*)	Lanceleaf tickseed (*Coreopsis lanceolata*)
Florida prairie clover (*Dalea carthagenensis*)	Twinflower (*Dyschoriste oblongifolia*)
Twinflower (*Dyschoriste oblongifolia*)	Rattlesnake master (*Eryngium yuccifolium*)
Elliott's lovegrass (*Eragrostis elliottii*)	Pineland heliotrope (*Euploca polyphylla*)
Snakeroot (*Eryngium yuccifolium*)	Yellowtop (*Flaveria linearis*)
Pineland heliotrope (*Euploca polyphylla*)	Slender blazing star (*Liatris gracilis*)
Yellowtop (*Flaveria linearis*)	Muhly grass (*Muhlenbergia capillaris*)
Spiked blazing star (*Liatris spicata*)	Havana skullcap (*Scutellaria havanensis*)
Muhly grass (*Muhlenbergia capillaris*)	Dropseed (*Sporobolus junceus*)
Scarlet sage (*Salvia coccinea*)	Blue porterweed (*Stachytarpheta jamaicensis*)
Havana skullcap (*Scutellaria havanensis*)	Bluecurls (*Trichostema dichotomum*)
Blue porterweed (*Stachytarpheta jamaicensis*)	Gamagrass (*Tripsacum dactyloides*)
Gamagrass (*Tripsacum dactyloides*)	

Lessons Learned

- Constructing a meadow in small, deliberate phases is a very effective approach because it allows for adjustments based on experience with the site.
- Certain species may be turn out to be too aggressive in small meadows, in this case scarlet sage (*Salvia coccinea*) and prairie clover (Dalea feayi) became problematic.
- Applying empirical lessons to guide adaptive management techniques can greatly aid in the success of a meadow.

THE FUTURE OF FLORIDA MEADOWS

Meadows are experiencing a surge in popularity across the United States, with Florida witnessing a particularly notable increase in interest. However, several significant challenges continue to impede the growth of meadow installations statewide. These challenges necessitate a multipronged approach encompassing increased research and funding at both state and federal levels to enhance our understanding of meadow ecology and best management practices.

Furthermore, landscape architects and managers must continue to embrace innovation, with industry leaders actively seeking and implementing new technologies, techniques, and approaches for meadow establishment and maintenance. Equally crucial is the development and training of a new generation of professionals who are eager to explore novel ideas and undertake the unique challenges presented by meadow projects. This requires fostering a culture of innovation and encouraging young professionals to embrace the complexities and rewards of working with these dynamic ecosystems. Ultimately, overcoming these hurdles necessitates a collaborative effort involving researchers, policymakers, industry professionals, and the broader community. By fostering strong partnerships and open communication, we can collectively advance the field of meadow construction and contribute to a more sustainable and resilient landscape future.

Site Preparation Challenges

The eradication of invasive species and the removal of nonnative turfgrasses, such as bermudagrass, from potential meadow sites presents a significant hurdle in meadow establishment. Overcoming the persistent weed seed bank,

A residential meadow in Central Florida blends seamlessly into its suburban surroundings. Photo by Andrea England with My Florida Meadow Company.

A residential meadow in central Florida featuring a lush variety of native plants, providing habitat for pollinators and visual appeal. Photo by Andrea England with My Florida Meadow Company.

a reservoir of dormant weed seeds within the soil, is particularly challenging, especially on sites with a history of agricultural use. These sites often harbor a high density of weed seeds, making it difficult to establish a desired plant community. Innovative approaches are urgently needed to effectively address this challenge at scale. These may involve developing novel combinations of existing site-preparation methods or exploring entirely new techniques. For example, research into biological control methods, such as the use of beneficial insects or fungi to suppress weed populations, could offer promising avenues. Furthermore, integrating these methods with careful site selection and rigorous weed monitoring is crucial for long-term success.

PLANT MATERIAL CHALLENGES

The deficiency of readily available and affordable native seed and plant materials poses the most significant obstacle to the widespread establishment of successful meadows in Florida. Obtaining high-quality, locally sourced seed mixes is often a major challenge. Many commercially available seed blends contain nonnative and sometimes even invasive species, while those that do feature native species may be overpriced or difficult to obtain. Furthermore, the range of native species available for direct seeding or transplanting is limited, particularly for less common species. This lack of access to diverse and appropriate plant materials can significantly increase the costs of meadow establishment, hinder project success, and ultimately limit the ecological value of the resulting meadow. The best way we can solve this issue is by supporting programs that aim to increase the supply of native seeds in Florida (e.g., Florida Native Seed Partnership), as well as continuing to support reputable companies that excel at the installation and maintenance of high-quality meadows.

Maintenance Knowledge Challenges

The limited number of skilled professionals with expertise in meadow construction and maintenance poses a significant barrier to the widespread adoption of this practice. While interest in meadows is growing, there is a critical shortage of individuals with the necessary knowledge and experience to design, install, and manage these complex ecosystems. This lack of expertise

By installing meadows, you help support the native nursery industry in Florida. Photo by Isabella Guttuso Browne.

By installing meadows, you help support the native seed industry in Florida. This is a professionally managed seed field. Photo by Joe Reams, Southern Habitats.

has several aspects. First, many landscape contractors and ecologists lack the specific training in and understanding of native plant communities, soil science, and ecological principles required for successful meadow establishment. Second, there is a dearth of specialized meadow maintenance professionals, such as those skilled in prescribed burning, targeted weed control, and monitoring of plant community dynamics. This shortage of skilled professionals can lead to poorly designed and maintained meadows that fail to achieve their objectives, ultimately hindering the progress of the cultural movement toward meadows. This issue is compounded by the fact that maintenance costs are lower for meadows than for conventional landscaping, making the long-term business model for meadow landscapers more challenging.

Cultural Acceptance Challenges

HOAs frequently have strict rules regarding lawn maintenance, including mowing height and the presence of weeds. Meadows, with their taller grasses and variety of plant species, may violate these rules. Concerns may arise about perceived untidiness, potential for attracting wildlife around community members with wildlife phobias, and perceived devaluation of property values within the neighborhood. These opposing perspectives can lead to conflicts between homeowners who wish to create more environmentally friendly and biodiverse landscapes and HOAs seeking to maintain a uniform aesthetic within the community. Solving this issue requires community leaders to rethink policies that may restrict meadows being installed.

A meadow at Bok Tower Gardens, near Lake Wales, Florida, showcases colorful wildflower blooms and native grasses, creating a biodiverse landscape. Photo by Nancy Bissett, The Natives.

Another view of the meadow at Bok Tower Gardens near Lake Wales, Florida shows a diverse range of grasses, flowering plants, and trees. Photo by Nancy Bissett, The Natives.

Concluding Remarks

La Florida, meaning "the land of flowers," was named for its abundant blossoms and natural beauty. Yet today many ask, Where have all the flowers gone? The landscapes that once teemed with vibrant wildflowers and diverse wildlife are increasingly giving way to urban sprawl, intensive agriculture, and invasive species. As a result, we are living in a world with fewer wild spaces, diminished biodiversity, and a growing disconnect from the natural environment that sustains us.

Meadows offer a powerful and tangible way to reverse these trends. They provide a haven for pollinators, birds, and countless other species while transforming the way we interact with and manage the land around us. Beyond their ecological benefits, meadows inspire a renewed sense of wonder, connecting us to the rhythms and resilience of nature. Each flower that blooms in a restored meadow symbolizes hope and possibility, reminding us that regeneration is always within reach.

This landscape design shows how multiple small meadows can be incorporated into a residential community. Just a small number of meadows can make a difference. Even limited meadow spaces can contribute to local biodiversity and ecosystem health. Illustration copyright © 2025 by Kelly Quinn | Canvas of the Wild. All rights reserved.

A meadow along a state highway near Cedar Key, Florida, features vibrant native wildflowers and grasses that enhance the area's ecological and scenic value. Photo by Gage LaPierre.

Meadows are more than just landscapes; they are living symbols of resilience and renewal, inviting each of us to act for a healthier planet. Whether you own a small patch of land or simply advocate for change in your community, creating or supporting meadows is a meaningful step toward restoring biodiversity and reconnecting with nature. Plant native wildflowers, reduce lawn space, or work with local schools, parks, and neighborhoods to transform underutilized areas into vibrant habitats. Every meadow, no matter its size, contributes to a larger network of ecosystems that support pollinators, birds, and wildlife. By acting, we are not just planting seeds in the soil—we are planting seeds of hope, community, and a more sustainable future. Let's reclaim the land of flowers, one meadow at a time.

appendix a

NATIVE RUDERAL SPECIES

Plants listed here are native species found in Florida that are generally considered ruderal, meaning they are relatively common early successional species that establish in areas following severe disturbances such as tilling from agriculture or clearing land from construction. While native and potentially beneficial in some cases, use these species judiciously, if at all, and control them if they become excessively abundant.

Species

Achillea gracilis
Aeschynomene americana var. *americana*
Aeschynomene indica
Allium canadense
Ambrosia artemisiifolia
Andropogon cretaceus
Andropogon tenuispatheus
Andropogon virginicus var. *virginicus*
Argemone albiflora var. *albiflora*
Axonopus compressus
Axonopus fissifolius
Baccharis halimifolia
Bacopa monnieri
Bidens alba
Boerhavia diffusa
Boltonia diffusa var. *diffusa*
Bulbostylis coarctata
Bulbostylis stenophylla
Carex longii
Cenchrus echinatus
Cenchrus incertus
Cenchrus longispinus
Cenchrus myosuroides
Cenchrus tribuloides
Centunculus minimus
Chenopodium album
Chenopodium berlandieri var. *boscianum*
Cirsium horridulum var. *vittatum*
Commelina diffusa
Cyperus brevifolius
Cyperus compressus
Cyperus croceus
Cyperus esculentus var. *leptostachyus*
Cyperus esculentus var. *macrostachyus*
Cyperus flavescens
Cyperus haspan
Cyperus hortensis

Cyperus lecontei
Cyperus ligularis
Cyperus neotropicalis
Cyperus odoratus var. *odoratus*
Cyperus pseudovegetus
Cyperus sesquiflorus
Cyperus subsquarrosus
Cyperus surinamensis
Daucus pusillus
Descurainia pinnata var. *pinnata*
Dichondra carolinensis
Digitaria ciliaris
Drymaria cordata var. *cordata*
Dysphania anthelmintica
Eclipta prostrata
Eleocharis flavescens var. *flavescens*
Eleocharis microcarpa var. *microcarpa*
Eleocharis nigrescens
Eleocharis vivipara
Eragrostis elliottii
Eragrostis spectabilis
Erechtites hieraciifolius
Erigeron annuus
Erigeron canadensis
Erigeron pusillus
Eupatorium capillifolium
Eupatorium compositifolium
Euphorbia cyathophora
Euphorbia heterophylla
Euphorbia hyssopifolia
Euphorbia maculata
Euphorbia polygonifolia
Galium aparine
Galium bermudense
Gamochaeta argyrinea
Geranium carolinianum
Hexasepalum teres
Hordeum pusillum
Ipomoea alba
Ipomoea cordatotriloba var. *cordatotriloba*
Ipomoea hederifolia
Ipomoea indica var. *acuminata*
Juncus bufonius
Juncus effusus ssp. *solutus*
Juncus tenuis
Lachnanthes caroliniana
Lepidium virginicum var. *virginicum*
Linaria floridana
Morella cerifera
Muscadinia rotundifolia var. *rotundifolia*
Oenothera filipes
Oenothera laciniata
Oenothera nutans
Oenothera simulans
Oldenlandia boscii
Oxalis dillenii
Oxalis stricta
Panicum dichotomiflorum var. *bartowense*
Panicum dichotomiflorum var. *dichotomiflorum*
Parietaria floridana
Parietaria praetermissa
Parthenocissus quinquefolia
Paspalum conjugatum
Paspalum setaceum
Passiflora incarnata
Physalis cordata
Physalis pubescens
Phytolacca rigida
Pinus taeda

Plantago rugelii
Portulaca nicaraguensis
Portulaca pilosa
Quercus hemisphaerica
Quercus nigra
Rhynchosia minima
Rhynchospora nitens
Rubus cuneifolius
Rubus pensilvanicus
Rumex hastatulus
Sagina decumbens
Salsola kali var. *caroliniana*
Salvia lyrata
Schoenoplectus californicus
Schoenoplectus tabernaemontani
Scoparia dulcis
Sesbania vesicaria
Setaria corrugata
Silene antirrhina
Sisyrinchium miamiense
Smilax bona-nox var. *bona-nox*
Solanum americanum
Solanum carolinense var. *carolinense*
Solanum nigrescens
Spergularia marina
Spermacoce glabra
Spermacoce ocymoides
Spermacoce remota
Sporobolus indicus
Sporobolus jacquemontii
Stachys floridana
Symphyotrichum bahamense
Symphyotrichum subulatum
Vicia acutifolia
Viola rafinesquei
Xanthium chinense
Xanthium orientale
Xyris jupicai

appendix b

SEED MIXTURES

Table B.1 lists a simple seed mix for full sun to partial shade mesic to xeric sites in North and Central Florida. For South Florida substitute 1.3 ounces of manyflowered beardtongue (*Penstemon multiflorus*) and 1.8 ounces of summer farewell (*Dalea pinnata*) for the *Coreopsis* spp. It should be possible to get Florida ecotype seed for all the native species listed. Plant 6–8 pounds PLS per acre (40–60 PLS seed/sq. ft.).

TABLE B.1. Seed mix for mesic to xeric sites in North and Central Florida

Common Name	*Scientific Name*	*% Seed in Mix*	*Oz./ lb. of Mix*	*Flowering Period*				*Life Cycle*	*Light Requirement*
				SPR	SUM	FALL	WIN		
Goldenmane tickseed	*Coreopsis basalis*	10.0	1.0	X	X			A	Full sun
Lanceleaf tickseed	*Coreopsis lanceolata*	10.0	2.4	0	X	X		P	Sun/partial shade
Black-eyed Susan	*Rudbeckia hirta*	10.0	0.3	X	X	0		P	Partial/ full sun
Blanket flower	*Gaillardia pulchella*	5.0	1.2	X	X	X		A/P	Partial/ full sun
Spotted beebalm	*Monarda punctata*	5.0	0.2	0	X	X		P	Partial/ full sun
Partridge pea	*Chamaecrista fasciculata*	10.0	8.0	0	X	X		A	Partial/ full sun
Purple lovegrass	*Eragrostis spectabilis*	25.0	1.3	0	0	0		P	Full sun
Broomsedge bluestem	*Andropogon virginicus*	25.0	1.6	0	0	0		P	Partial/ full sun

Notes: A = annual; P = perennial

Table B.2 lists a simple mix for full sun to partial shade hydric to mesic sites in North and Central Florida. For South Florida substitute 1.1 ounces manyflowered beardtoungue (*Penstemon multiflorus*) for the *Coreopsis basalis*. It should be possible to get Florida ecotype seed for all species listed. Plant 6–9 pounds PLS per acre (40–60 PLS seed/sq. ft.).

TABLE B.2. Seed mix for hydric to mesic sites in North and Central Florida

Common Name	*Scientific Name*	*% Seed in Mix*	*Oz./ lb. of Mix*	*Flowering Period*				*Life Cycle*	*Light Requirement*
				SPR	SUM	FALL	WIN		
Goldenmane tickseed	*Coreopsis basalis*	10.0	1.2	X	X			A	Full sun
Leavenworth's coreopsis	*Coreopsis leavenworthii*	10.0	0.2		X	X	0	P	Full sun
Black-eyed Susan	*Rudbeckia hirta*	10.0	0.4	X	X			P	Partial/ full sun
Blanket flower	*Gaillardia pulchella*	5.0	1.3	X	X	X		A/P	Partial/ full sun
Spotted beebalm	*Monarda punctata*	5.0	0.2		X	X		P	Partial/ full sun
Partridge pea	*Chamae-crista fascic-ulata*	10.0	9.3		X	X		A	Partial/ full sun
Elliott's lovegrass	*Eragrostis elliottii*	25.0	1.1					P	Full sun
Purple bluestem	*Andropogon glaucopsis*	25.0	5.6					P	Full sun

Notes: A: annual; P: perennial

appendix c

WEED GUIDE

Common Weeds

In Table C.1, threat level is defined by four levels:

1 = Low threat; has some potential to become an issue if conditions are right but should disappear over time if controlled by mowing, hand-weeding, or fire.
2 = Moderate threat; can become an issue if the population achieves high coverage. Mowing or fire treatments may resolve the issue over time, but spot herbicide treatment is likely needed.
3 = High level threat; typically an issue in most meadows, fire and mowing are unlikely to control, so control via herbicide is recommended.
4 = Extreme level threat; implement control measures immediately. Herbicide is the only option, and control measures may entail sacrificing areas of the meadow.

TABLE C.1. Fifty common and problematic weeds for meadows in Florida

Scientific Name	*Common Name*	*Problematic Characteristics*	*Threat Level*	*Control Recommendation*
***Amaranthus* spp.**	Spiny pigweed	Warm-season annuals with tall growth habit produce abundant seed; some have spines.	2	Mow to prevent seed set; if number of plants is limited, consider hand-weeding or spot-spraying.
***Ambrosia* spp.**	Ragweed	Reseeding annuals; tall growth height, associated with dilapidation.	3	Mow to prevent seed set; if number of plants is limited, consider hand-weeding or spot-spraying. For larger populations and if height differential between meadow seedlings allows, wick with labeled, nonselective herbicides.

Scientific Name	*Common Name*	*Problematic Characteristics*	*Threat Level*	*Control Recommendation*
Anthemis spp.	Mayweed	May overwhelm seedlings on certain sites.	1	Spot-spray.
Artemisia spp.	Mugwort	Extensive root system that will outcompete other plants, especially on sites with productive soils. Very difficult to kill, even with herbicide.	4	Spot-spray with triclopyr.
Axonopus spp.	Carpetgrass	Spreading growth habit.	1	Spot-spray.
Bidens alba	Spanish needles	Good nectar plant but aggressive; produces abundant seed that is easily dispersed, can develop adventitious roots on stem; tall growth height.	2	Spot-spray. For larger populations and if height differential between meadow seedlings allows, wick with labeled, nonselective herbicides.
Brassica spp.	Wild mustards, cabbages, turnips	Problematic during establishment phase; reseeds rapidly.	1	Mow to prevent seed set. If number of plants is limited, consider hand-weeding or spot-spraying.
Cardamine hirsuta	Hairy bittercress	Problematic during establishment phase; reseeds rapidly.	1	Mow to prevent seed set; if number of plants is limited, consider hand-weeding or spot-spraying.
Cenchrus spp.	Sandbur	Likes bare ground; spreads readily; spiny seeds are unpleasant for foot travel.	3	Make big effort to treat before burs mature. If number of plants is limited, consider hand-weeding or spot-spraying. Imazapic formulations labeled for use in wildflower planting may give selective control.
Croton setiger	Doveweed	Germinates in summer; can crowd out other species during early establishment.	1	Hand-weed; spot-spray.
Cynodon dactylon	Bermudagrass	Perennial grass; common type persists via seed, forage types are vegetative only; extremely aggressive due to stolons or rhizomes.	4	If number of plants is limited, consider hand-weeding or spot-spraying. Imazapic formulations labeled for use in wildflower planting may give selective control.

Scientific Name	*Common Name*	*Problematic Characteristics*	*Threat Level*	*Control Recommendation*
Cyperus esculentus and C. rotundus	Yellow and purple nutsedge	Persist via rhizomes and underground tubers or "nuts"; can outcompete seedlings. If possible, avoid sites with nutsedge present.	3	If number of plants is limited, consider hand-weeding when seedlings first emerge or spot-spraying. Imazapic formulations labeled for use in wildflower planting may give selective control.
Descurainia pinnata	Western tansymustard	Cool-season annual; early pioneer weed. Generally not problematic.	1	Hand-weed; spot-spray.
Desmodium incanum	Creeping beggarweed	Stoloniferous perennial; spreads via stolons and sticky seeds.	2	Spot-spray.
Dichondra spp.	Ponysfoot	Perennial that spreads by stems that root at the nodes; may become problematic if meadow seed fails to establish.	1	Reduce irrigation frequency.
Digitaria spp.	Crabgrass	Some species are annuals and some are perennials; stoloniferous or decumbent rooting at nodes; reseeds quickly, weedy appearance.	4	If limited number of plants, consider hand-weeding when first emerge or spot-spraying. Imazapic formulations labeled for use in wildflower planting may give selective control.
Diodia virginiana	Virginia buttonweed	Perennial; likes fairly wet sites; can outcompete meadow seedlings during establishment phase; reseeds rapidly.	1	Mow to prevent seed set. If number of plants is limited, consider hand-weeding or spot-spraying.
Drymaria cordata	Tropical chickweed	Abundant, sticky fruit and seed; spreads easily by seed and pieces; low-growing, problematic if dense.	1	Do not hand-weed; reduce irrigation frequency; spot-spray.
Eleusine indica	Goosegrass	Aggressive annual grass; persistent due to abundant seed.	2	If number of plants is limited, consider hand-weeding or spot-spraying. Imazapic formulations labeled for use in wildflower planting may give selective control.

Scientific Name	*Common Name*	*Problematic Characteristics*	*Threat Level*	*Control Recommendation*
Erechtites hieraciifolius	American burnweed; fireweed	Winter annual; likes disturbed, fairly wet sites; weedy appearance.	1	Remove by mowing before seeds mature; if number of plants is limited, consider hand-weeding or spot-spraying.
Eremochloa ophiuroides	Centipedegrass	Stoloniferous perennial grass; spreads vegetatively; aggressive.	4	Control or eradicate if possible by spot-spraying.
Erigeron canadensis	Canadian horseweed	Warm-season annual or biannual; weedy appearance; fairly easy to pull up when small.	1	Mow to prevent seed set. If number of plants is limited, consider hand-weeding or spot-spraying. For larger populations and if height differential between meadow seedlings allows, wick with labeled, nonselective herbicides.
Eupatorium capillifolium	Dogfennel	Tall growth habit; persistent perennial from rootstock and also spreads by seeds; weedy appearance.	4	Spot-spraying. For larger populations and if height differential between meadow seedlings allows, wick with labeled, nonselective herbicides.
Geranium carolinianum	Carolina geranium; Carolina cranesbill	Winter annual; can be problematic in establishment phase if left unchecked.	1	Mowing may reduce seed set and provide tidier appearance.
***Hydrocotyle* spp.**	Marsh pennywort; dollarweed	Perennial spreads by creeping stems and seed; can form dense mat and be problematic in fairly wet locations.	1	Reduce irrigation frequency.
Hypochaeris radicata	Hairy catsear; false dandelion	Mildly problematic during establishment phase.	1	Do not allow to go to seed. Mow; hand-weed; spot-spray.
Indigofera hirsuta	Hairy indigo	Warm-season annual or short-lived perennial; sub-shrub; abundant hard seed.	4	If possible maintain stale seedbed to avoid bringing buried seed to surface. Do not allow seeds to mature. Plants under 2 ft. tall are relatively easy to pull up. Mowing will weaken but not completely suppress seed production. Spot-spray.

Scientific Name	*Common Name*	*Problematic Characteristics*	*Threat Level*	*Control Recommendation*
Indigofera spicata	Trailing indigo	Warm-season spreading annual; toxic to most mammals; can crowd out other species.	4	Mowing ineffective due to low growth height; spot-spray or hand-weed.
Lamium amplexicaule	Henbit; henbit deadnettle	Winter annual; can self-pollinate; likes disturbed sites; can be problematic during establishment phase.	1	Mowing may reduce seed set and provide tidier appearance.
Lepidium spp.	Peppercress	Winter annual; likes disturbed sites; can be problematic during establishment phase.	1	Mowing may reduce seed set and provide tidier appearance.
Lolium perenne (= *L. perenne* spp. *multiflorum*; = *L. multiflorum*)	Italian or annual ryegrass	Cool-season annual; can volunteer from seed of previous year; is thought to be allelopathic, active plant growth or residue may harm germination of seeded species.	2	Hand-weed; spot-spray. Imazapic formulations labeled for use in wildflower planting may give selective control.
Medicago lupulina	Black medic	Cool-season annual; hard seeded; present on many old pasture or lawn sites; can overtake seedlings in early establishment phase.	1	Usually grows too low for mowing to affect; ignore.
Mollugo verticillata	Indian chickweed; green carpetweed	Warm-season annual; early pioneer weed; can form dense mat and smother other plants.	1	Easily hand-weeded to prevent seeding. Reduce irrigation frequency. Spot-spray.
Paspalum dilatatum	Dallisgrass	Warm-season perennial.	4	Eradicate prior to meadow planting. If number of plants is limited, consider hand-weeding or spot-spraying. Imazapic formulations labeled for use in wildflower planting may give selective control.
Paspalum notatum	Bahiagrass	Warm-season perennial; forms dense monocultures.	4	Eradicate prior to meadow planting. If number of plants is limited, consider hand-weeding or spot-spraying. Imazapic formulations labeled for use in wildflower planting may give selective control.

Scientific Name	*Common Name*	*Problematic Characteristics*	*Threat Level*	*Control Recommendation*
Paspalum setaceum	Thin paspalum; bull paspalum	Warm-season perennial; cespitose or short rhizomes; weedy appearance; only harmful if dense patches emerge.	1	Hand-weed; spot-spray.
Phyllanthus **spp.**	Leaf flower; chamberbitter; long-stalked phyllanthus	Warm-season annuals that persist from abundant seed; need moisture and light to germinate, so should decline as meadow matures.	1	Prevent seeding if possible. Hand-weed; spot-spray.
Phytolacca americana	Pokeweed	Tall perennial; has some value as pollinator and wildlife plant; weedy appearance.	1	Easy to hand-weed when small; spot-spray.
Plantago **spp.**	Plantain	Can overtake species in early establishment phase.	1	Dig up before flowers appear; hand-weed; spot-spray.
Portulaca oleracea	Purslane	Low-growing annual; reproduces by seed and stem fragments; can crowd out meadow seedlings.	4	Mowing will not control. Spot-spraying needed.
Raphanus raphanistrum	Wild radish	Early pioneer weed in winter months, generally not problematic.	1	Spot-spray if needed.
Richardia **spp.**	Mexican clover; Brazilian pusley	Annual or perennial; all spread by seed or vegetative fragments; low-growing, can crowd out annuals if dominant.	3	Hand-weed; spot-spray.
Soliva sessilis	Lawn burweed	Cool-season annual; early pioneer weed. Generally not problematic.	1	If known to be a problem, fall/winter fallowing is important in site preparation. Repeat spraying of emerging seedlings is needed. Do not allow to go to seed.
Stachys floridana	Florida betony	Native but can become problematic in residential meadows due to its dense growth habit and tough-to-kill tubers.	1	Spot-spray.

Scientific Name	*Common Name*	*Problematic Characteristics*	*Threat Level*	*Control Recommendation*
Stellaria media	Chickweed	Low-growing; problematic if dense.	1	Hand-weed; reduce irrigation frequency; spot-spray.
Taraxacum officinale	Dandelion	Common; weedy appearance but generally not problematic long term.	1	Mowing may reduce seed set and provide tidier appearance.
Trifolium repens	White clover	Common in residential areas.	1	Control if dominant. Reduce irrigation; spot-spray.
Urtica chamaedryoides	Heartleaf nettle; fireweed	Indeterminant winter annual; shade tolerant; needs bare ground; very shallow roots; stinging hairs on leaves cause skin irritation.	1	Control if undesired. Do not mow. Wear gloves if hand-weeding; spot-spray.
Verbena brasiliensis	Brazilian vervain	Tall, aggressive perennial; spread by seed.	3	Control or eradicate if possible. Do not allow to go to seed. Mowing may suppress seed but will not eliminate; hand-weed or spot-spray.
Youngia japonica	Asiatic hawksbeard	Tap-rooted annual with wind-dispersed seed; mostly spring flowers but can flower year-round; generally not problematic long term.	1	Important to prevent seed development. Mowing is ineffective for control; hand-weed or spot-spray
***Zoysia* spp.**	Zoysiagrass	Stoloniferous perennial grass; spreads vegetatively; aggressive.	4	Control or eradicate if possible by spot-spraying.

Invasive Species

All species listed in Table C.2 should be eliminated prior to or following a meadow installation. In most cases herbicide treatments will be needed because mechanical treatments alone will not control these species. For information on identifying and treating invasive species, see the Florida Natural Area Inventory (FNAI) Invasive Plant Directory (https://www.fnai.org/species-communities/invasives/invasives) or the University of Florida Center for Aquatic and Invasive Plants website (https://plants.ifas.ufl.edu/).

TABLE C.2. Invasive species listed by Florida Invasive Species Council

Scientific Name	*Common Name*
Abrus precatorius	Rosary pea
Acacia auriculiformis	Earleaf acacia
Acalypha alopecuroidea	Foxtail copperleaf
Adenanthera pavonina	Red sandalwood
Agave sisalana	Sisal hemp
Albizia julibrissin	Mimosa tree
Albizia lebbeck	Woman's tongue tree
Alstonia macrophylla	Devil tree
Antigonon leptopus	Coral vine
Ardisia crenata	Coral ardisia
Ardisia elliptica	Shoebutton ardisia
Ardisia japonica	Japanese ardisia
Aristolochia elegans	Elegant Dutchman's pipe
Asparagus aethiopicus	Sprenger's asparagus-fern
Asystasia gangetica	Chinese violet
Bauhinia variegata	Orchid tree
Begonia cucullata	Wax begonia
Bischofia javanica	Javanese bishopwood
Broussonetia papyrifera	Paper mulberry
Callisia fragrans	Inch plant
Callistemon viminalis	Bottlebrush
Calophyllum antillanum	Antilles calophyllum
Casuarina equisetifolia	Australian pine
Casuarina glauca	Gray she oak

Scientific Name	*Common Name*
Cenchrus polystachios	Mission grass
Cenchrus purpureus	Elephant grass
Cenchrus setaceus	Crimson fountaingrass
Cestrum diurnum	Dayflowering jessamine
Chamaedorea seifrizii	Seifriz's chamaedorea
Cinnamomum camphora	Camphor tree
Clematis terniflora	Sweet autumn virginsbower
Colocasia esculenta	Coco yam, wild taro
Colubrina asiatica	Latherleaf
Crassocephalum crepidioides	Redflower ragleaf
Cryptostegia madagascariensis	Madagascar rubbervine
Cupaniopsis anacardioides	Carrotwood
Cyperus involucratus	Umbrella plant
Cyperus prolifer	Miniature flatsedge
Dactyloctenium aegyptium	Crowfoot grass
Dalbergia sissoo	Indian rosewood
Dalechampia scandens	Spurgecreeper
Deparia kaalaana	Serpent fern
Deparia petersenii	Japanese false spleenwort
Dioscorea alata	Winged yam
Dioscorea bulbifera	Air-potato
Eugenia uniflora	Surinam cherry
Eulophia graminea	Eulophia ground orchid
Ficus microcarpa	Laurel fig
Flacourtia indica	Governor's plum
Heteropterys brachiata	Beechey's Withe
Hyparrhenia rufa	Jaragua grass
Imperata cylindrica	Cogongrass
Jasminum dichotomum	Gold Coast jasmine
Jasminum fluminense	Brazilian jasmine
Kalanchoe pinnata	Cathedral bells
Kalanchoe × houghtonii	Hybrid mother-of-millions
Koelreuteria elegans	Flamegold
Lantana strigocamara	Lantana
Leucaena leucocephala	White leadtree
Ligustrum lucidum	Glossy privet
Ligustrum sinense	Chinese privet

Scientific Name	Common Name
Livistona chinensis	Fountain palm
Lonicera japonica	Japanese honeysuckle
Ludwigia peruviana	Peruvian primrose-willow
Lygodium japonicum	Japanese climbing fern
Lygodium microphyllum	Old World climbing fern
Macfadyena unguis-cati	Catclaw-Vine
Macroptilium lathyroides	Wild bushbean
Manilkara zapota	Sapodilla
Megathyrsus maximus	Guinea grass
Melaleuca quinquenervia	Melaleuca
Melia azedarach	Chinaberry
Melinis minutiflora	Molasses grass
Melinis repens	Natal grass
Merremia tuberosa	Spanish arborvine
Microstegium vimineum	Japanese stilt grass
Mikania micrantha	Mile-a-minute
Mimosa pigra	Catclaw mimosa
Momordica charantia	Balsamapple
Murraya paniculata	Orange jessamine
Nandina domestica	Heavenly bamboo
Nephrolepis brownii	Asian sword fern
Nephrolepis cordifolia	Narrow sword fern
Neyraudia reynaudiana	Burma reed
Nymphoides cristata	Crested floatingheart
Paederia cruddasiana	Sewervine
Paederia foetida	Skunkvine
Panicum repens	Torpedograss
Passiflora biflora	Twoflower passionflower
Phoenix reclinata	Senegal date palm
Phyllostachys aurea	Golden bamboo
Pittosporum pentandrum	Taiwanese cheesewood
Platycerium bifurcatum	Elkhorn fern
Praxelis clematidea	Praxelis
Psidium cattleianum	Strawberry guava
Psidium guajava	Guava
Pteris vittata	Chinese brake fern
Ptychosperma elegans	Ptychosperma

Scientific Name	*Common Name*
Pueraria montana var. *lobata*	Kudzu
Rhodomyrtus tomentosa	Downy rose-myrtle
Richardia grandiflora	Largeflower Mexican clover
Ricinus communis	Castor bean
Rotala rotundifolia	Roundleaf toothcup
Ruellia blechum	Browne's blechum
Ruellia simplex	Mexican petunia
Sansevieria hyacinthoides	Iguanatail
Schinus terebinthifolius	Brazilian pepper tree
Senna pendula var. *glabrata*	Valamuerto
Sesbania punicea	Red sesbania
Sida planicaulis	Sida
Solanum diphyllum	Twoleaf nightshade
Solanum tampicense	Wetland nightshade
Solanum torvum	Turkey berry; susumber
Solanum viarum	Tropical soda apple
Spermacoce verticillata	Shrubby false buttonweed
Sphagneticola trilobata	Bay Biscayne creeping-oxeye
Sporobolus indicus	West Indian dropseed
Stachytarpheta cayennensis	Cayenne porterweed
Syngonium podophyllum	American evergreen
Tectaria incisa	Incised halberd fern
Terminalia catappa	Tropical almond
Terminalia muelleri	Australian almond
Thespesia populnea	Seaside mahoe
Tradescantia fluminensis	White-flowered spiderwort; white-flowered wandering Jew
Tradescantia spathacea	Boat lily
Triadica sebifera	Chinese tallow tree
Tribulus cistoides	Jamaica feverplant
Urena lobata	Caesarweed
Urochloa mutica	Para grass
Vitex rotundifolia	Beach virtex
Vitex trifolia	Simpleleaf chastetree
Wisteria sinensis	Chinese wisteria
Xanthosoma sagittifolium	Arrowleaf elephant's ear

NOTES

one The Meadow Concept

1 Robbins, *Lawn People*; Jenkins, *The Lawn*.
2 Steinberg, *American Green*; Jenkins, *The Lawn*.
3 Milesi et al., "Mapping and Modeling."
4 Steinberg, *American Green*; Jenkins, *The Lawn*; Milesi et al., "Mapping and Modeling."
5 Florida Natural Areas Inventory, "Guide to the Natural Communities of Florida."
6 Noss, *Forgotten Grasslands of the South*.
7 Florida Natural Areas Inventory, "Guide to the Natural Communities of Florida."
8 Gerwing et al., "Restoration, Reclamation, and Rehabilitation."
9 Margolies, "Corporate Landscaping Lets Its Hair Down"; Sturm and Frischie, "Mid-Atlantic Native Meadows Guidelines"; Neal, "Planting for Pollinators"; Gordon et al., "Creating a Wildflower Meadow"; Huegel, "Planting a Wildflower Meadow."
10 Margolies, "Corporate Landscaping Lets Its Hair Down"; Sturm and Frischie, "Mid-Atlantic Native Meadows Guidelines"; Huegel, "Planting a Wildflower Meadow."
11 Sturm and Frischie, "Mid-Atlantic Native Meadows Guidelines"; Neal, "Planting for Pollinators"; Gordon et al., "Creating a Wildflower Meadow"; Huegel, "Planting a Wildflower Meadow."
12 Sturm and Frischie, "Mid-Atlantic Native Meadows Guidelines."
13 Margolies, "Corporate Landscaping Lets Its Hair Down"; Sturm and Frischie, "Mid-Atlantic Native Meadows Guidelines"; Gordon et al., "Creating a Wildflower Meadow"; Huegel, "Planting a Wildflower Meadow."
14 Sturm and Frischie, "Mid-Atlantic Native Meadows Guidelines."
15 Margolies, "Corporate Landscaping Lets Its Hair Down"; Sturm and Frischie, "Mid-Atlantic Native Meadows Guidelines"; Huegel, "Planting a Wildflower Meadow."

16 Sturm and Frischie, "Mid-Atlantic Native Meadows Guidelines."
17 Sturm and Frischie, "Mid-Atlantic Native Meadows Guidelines."
18 Margolies, "Corporate Landscaping Lets Its Hair Down"; Sturm and Frischie, "Mid-Atlantic Native Meadows Guidelines"; Huegel, "Planting a Wildflower Meadow."
19 Sturm and Frischie, "Mid-Atlantic Native Meadows Guidelines."
20 Shober et al., "Management of Fertilizers and Water"; Haley et al., "Residential Irrigation Water Use."
21 Hasebroock, "The Price of Plenty"; Heisler et al., "Eutrophication and Harmful Algal Blooms."
22 Hasebroock, "The Price of Plenty"; Heisler et al., "Eutrophication and Harmful Algal Blooms"; Pimentel et al., "Update on the Environmental and Economic Costs."
23 Banks and McConnell, "National Emissions from Lawn and Garden Equipment"; Walker and Banks, "Two-Stroke Engines in Landscape Maintenance."
24 Koh et al., "Modeling the Status"; Kluser and Peduzzi, "Global Pollinator Decline"; National Research Council, *Status of Pollinators in North America.*
25 National Research Council, *Status of Pollinators in North America.*
26 Kautz, "Land Use and Land Cover"; Millsap et al., "Setting Priorities."
27 Sturm and Frischie, "Mid-Atlantic Native Meadows Guidelines"; Huegel, "Planting a Wildflower Meadow."
28 Sturm and Frischie, "Mid-Atlantic Native Meadows Guidelines."
29 Sturm and Frischie, "Mid-Atlantic Native Meadows Guidelines."
30 Koh et al., "Modeling the Status"; Kautz, "Land Use and Land Cover"; Millsap et al., "Setting Priorities."
31 Lindemann-Matthies and Matthies, "The Influence of Plant Species Richness"; Lindemann-Matthies et al., "The Influence of Plant Diversity on People's Perception."
32 Robbins, *Lawn People*; Jenkins, *The Lawn.*
33 Bormann et al., *Redesigning the American Lawn*; Ramer et al., "Exploring Park Visitor Perceptions."
34 Margolies, "Corporate Landscaping Lets Its Hair Down"; Sturm and Frischie, "Mid-Atlantic Native Meadows Guidelines"; Huegel, "Planting a Wildflower Meadow."
35 Sturm and Frischie, "Mid-Atlantic Native Meadows Guidelines."
36 Margolies, "Corporate Landscaping Lets its Hair Down"; Sturm and Frischie, "Mid-Atlantic Native Meadows Guidelines"; Huegel, "Planting a Wildflower Meadow."

two Preparing to Construct a Meadow

1 US Department of Agriculture, Natural Resources Conservation Service, "Web Soil Survey, https://websoilsurvey.nrcs.usda.gov/app/; University of California–Davis and NRCS, "SoilWeb" interactive map, https://casoilresource.lawr.ucdavis.edu/gmap/.

2 G. D. L. LaPierre, N. D. Medina-Irizarry, and M. G. Andreu, "Florida Soil Series and Natural Community Associations," University of Florida, School of Forest, Fisheries, and Geomatics Sciences Publication FOR384, 2022, https://edis.ifas.ufl.edu/publication/FR455.

3 "Aerial Photography: Florida," University of Florida Digital Collections, https://ufdc.ufl.edu/collections/aerials.

4 Florida Invasive Species Council, "2023 FISC List of Invasive Plant Species," https://www.floridainvasives.org/plant-list/2023-invasive-plant-species/.

5 Richard Wunderlin, Bruce Hansen, Alan Franck, and F. B. Essig, "Atlas of Florida Plants," Institute for Systematic Botany, University of South Florida, Tampa, 2025, https://florida.plantatlas.usf.edu/; Weakley et al., "Studies in the Vascular Flora."

6 UF/IFAS Soil Testing Services, "Analytical Services Laboratories (ANSERV Labs), https://soilslab.ifas.ufl.edu/. For soil, plant tissue, and water analysis tests, costs, and submission forms, see https://soilslab.ifas.ufl.edu/extension-soil-testing-laboratory/.

7 Mortellaro et al., "Coefficients of Conservatism Values."

8 Lee-Mäder et al., "Establishing Pollinator Meadows from Seed."

9 Weakley et al., "Studies in the Vascular Flora."

10 Sturm and Frischie, "Mid-Atlantic Native Meadows Guidelines"; Gordon et al., "Creating a Wildflower Meadow"; Huegel, "Planting a Wildflower Meadow."

three Installing a Meadow

1 Florida Natural Areas Inventory, "Guide to the Natural Communities of Florida"; Gerwing et al., "Restoration, Reclamation, and Rehabilitation"; Noss, *Forgotten Grasslands of the South.*

2 Margolies, "Corporate Landscaping Lets Its Hair Down."

3 Sturm and Frischie, "Mid-Atlantic Native Meadows Guidelines"; Neal, "Planting for Pollinators."

4 Gordon et al., "Creating a Wildflower Meadow"; Huegel, "Planting a Wildflower Meadow"; Shober et al., "Management of Fertilizers and Water"; Haley et al., "Residential Irrigation Water Use"; Hasebroock, "The Price of Plenty"; Heisler et al., "Eutrophication and Harmful Algal Blooms."

four Managing a Meadow

1 Sturm and Frischie, "Mid-Atlantic Native Meadows Guidelines."

2 Sturm and Frischie, "Mid-Atlantic Native Meadows Guidelines."

3 Sturm and Frischie, "Mid-Atlantic Native Meadows Guidelines."

4 https://www.regionalconservation.org. For South Florida, see https://www.regionalconservation.org/ircs/Coefficients%20of%20Conservatism%20Values%20for%20South%20Florida.pdf.

LITERATURE CITED

Banks, Jamie L., and Robert McConnell. "National Emissions from Lawn and Garden Equipment." US Environmental Protection Agency, 2015. https://www3.epa.gov/ttnchie1/conference/ei21/session10/banks.pdf.

Bormann, F. Herbert, Diana Balmori, Gordon T. Geballe, and Lisa Vernegaard. *Redesigning the American Lawn: A Search for Environmental Harmony*. Yale University Press, 2001.

Clemente, A. S., C. Werner, C. Máguas, M. S. Cabral, M. A. Martins-Loução, and O. Correia. "Restoration of a Limestone Quarry: Effect of Soil Amendments on the Establishment of Native Mediterranean Sclerophyllous Shrubs." *Restoration Ecology* 12 (2004): 20–28. https://doi.org/10.1111/j.1061-2971.2004.00256.x

Florida Natural Areas Inventory. "Guide to the Natural Communities of Florida: 2010 Edition." https://www.fnai.org/species-communities/natcom-guide.

Gerwing, Travis G., Virgil C. Hawkes, George D. Gann, and Stephen D. Murphy. "Restoration, Reclamation, and Rehabilitation: On the Need for, and Positing a Definition of, Ecological Reclamation." *Restoration Ecology* 30 (2022). https://doi.org/10.1111/rec.13461.

Gilman, Edward F. Effects of Amendments, Soil Additives, and Irrigation on Tree Survival and Growth. *Journal of Arboriculture* 30 (2004). https://doi.org/10.48044/jauf.2004.037.

Gordon, W., A. Hudson, L. McNamara, M. Wing, and T. Clem. 2019. "Creating a Wildflower Meadow." https://sfyl.ifas.ufl.edu/media/sfylifasufledu/alachua/images/pdf/Creating-a-Wildflower-Meadow.pdf

Haley, M. B., M. D. Dukes, and G. L. Miller. "Residential Irrigation Water Use in Central Florida." *Journal of Irrigation and Drainage Engineering* 133, no. 5 (2007): 423–510.

Hasebroock, Abigail. "The Price of Plenty: Fertilizer Flatline?" WUSF Public News Media, June 9, 2023. https://www.wusf.org/environment/2023-06-09/the-price-of-plenty-fertilizer-flatline.

Heisler, J., P. M. Glibert, J. M. Burkholder, D. M. Anderson, W. Cochlan, W. C. Dennison, Q. Dortch, C. J. Gobler, C. A. Heil, E. Humphries, A. Lewitus, R. Magnien, H. G. Marshall, K. Sellner, D. A. Stockwell, D. K. Stoecker, and M. Suddleson.

"Eutrophication and Harmful Algal Blooms: A Scientific Consensus." *Harmful Algae* 8, no. 1 (2008): 3–13. https://doi.org/10.1016/j.hal.2008.08.006.
Huegel, Craig N. "Planting a Wildflower Meadow." *Palmetto* 36, no. 2 (2020): 1–15. https://www.fnps.org/assets/pdf/palmetto/36-2_Huegel_Meadows_2.pdf.
Jenkins, Virginia Scott. *The Lawn: A History of an American Obsession*. Smithsonian Books, 1994.
Jordan, Sarah Foltz, Jessica Kay Cruz, Kelly Gill, Jennifer Hopwood, Jarrod Fowler, Eric Lee-Mäder, and Mace Vaughan. "Organic Site Preparation for Wildflower Establishment." Xerces Society for Invertebrate Conservation 48 (2018). https://xerces.org/sites/default/files/2018-05/16-027_02_XercesSoc_Organic-Site-Preparation-for-Wildflower-Establishment_web.pdf.
Kautz, Randy S. "Land Use and Land Cover Trends in Florida, 1936–1995." *Florida Scientist* 61, no. 3–4 (1998): 171–187. https://www.jstor.org/stable/24320971.
Kemery, Ricky D., and Michael N. Dana. "Influence of Container Size and Medium Amendment on Post-Transplant Growth of Prairie Perennial Seedlings." *Horttechnology* 11, no. 1 (2001): 52–56. https://doi.org/10.21273/HORTTECH.11.1.52.
Kluser, Stéphane, and Pascal Peduzzi. "Global Pollinator Decline: A Literature Review." UNEP/GRID-Europe, 2007. https://unepgrid.ch/storage/app/media/legacy/37/Global_pollinator_decline_literature_review_2007.pdf. https://unepgrid.ch/storage/app/media/legacy/37/Global_pollinator_decline_literature_review_2007.pdf.
Koh, I., R. P. Dicks, N. U. Kleijn, R. F. A. Potts, and L. L. Morandin. "Modeling the Status, Trends, and Impacts of Wild Bee Abundance in the United States." *Proceedings of the National Academy of Sciences of the United States of America* 113, no. 1 (2016): 140–45.
Lee-Mäder, Eric, Brianna Borders, and Ashley Minnerath. "Establishing Pollinator Meadows from Seed." Xerces Society for Invertebrate Conservation, 2013. https://xerces.org/sites/default/files/2018-05/15-020_02_XercesSoc_Establishing-Pollinator-Meadows-from-Seed_web.pdf.
Lindemann-Matthies, Petra, Xenia Junge, and Diethart Matthies. "The Influence of Plant Diversity on People's Perception and Aesthetic Appreciation of Grassland Vegetation." *Biological Conservation* 143 (2010): 195–202. https://doi.org/10.1016/j.biocon.2009.10.003.
Lindemann-Matthies, Petra, and Diethart Matthies. "The Influence of Plant Species Richness on Stress Recovery of Humans." *Web Ecology* 18, no. 1 (2018): 121–28. https://doi.org/10.5194/we-18-121-2018.
Margolies, Jane. "Corporate Landscaping Lets Its Hair Down." *New York Times*, June 6, 2023, 1–5.
Meissen, Justin C., Alec J. Glidden, Mark E. Sherrard, Kenneth J. Elgersma, and Laura L. Jackson. "Seed Mix Design and First Year Management Influence Mul-

tifunctionality and Cost-Effectiveness in Prairie Reconstruction." *Restoration Ecology* 28 (2020): 807–816. https://doi.org/10.1111/rec.13013.
Milesi, Cristina, Steven W. Running, Christopher D. Elvidge, John B. Dietz, Benjamin T. Tuttle, and Ramakrishna R. Nemani. "Mapping and Modeling the Biogeochemical Cycling of Turf Grasses in the United States." *Environmental Management* 36, no. 3 (2005): 426–438. https://doi.org/10.1007/s00267-004-0316-2
Millsap, Brian A., Jeffery A. Gore, Douglas E. Runde, and Susan I. Cerulean. "Setting Priorities for the Conservation of Fish and Wildlife Species in Florida." *Wildlife Monographs* 111 (July 1990): 3–57. https://www.jstor.org/stable/3830656.
Mortellaro, Steve, Mike Barry, George D. Gann, John Zahina, Sally Channon, Charles Hilsenbeck, Douglas Scofield, George Wilder, and Gerould Wilhelm. "Coefficients of Conservatism Values and the Floristic Quality Index for the Vascular Plants of South Florida." *Southeastern Naturalist* 11, no. 3 (2012): 1–52. https://doi.org/10.1656/058.011.m301.
National Research Council. *Status of Pollinators in North America*. National Academies Press, 2007.
Neal, Catherine. "Planting for Pollinators: Establishing a Wildflower Meadow from Seed." University of New Hampshire Extension fact sheet, 2020. https://extension.unh.edu/resource/planting-pollinators-establishing-wildflower-meadow-seed-fact-sheet.
Norcini, Jeffrey C., and James H. Aldrich. "Establishment of Native Wildflower Plantings by Seed." Historical Fact Sheets ENH968/EP227. April 2004. https://doi.org/10.32473/edis-ep227-2004.
Noss, Reed F. *Forgotten Grasslands of the South: Natural History and Conservation*. Island Press, 2012.
Pimentel, David, Rodolfo Zuniga, and Doug Morrison. "Update on the Environmental and Economic Costs Associated with Alien-Invasive Species in the United States." *Ecological Economics* 52, no. 3 (2005): 273–88. https://doi.org/10.1016/j.ecolecon.2004.10.002.
Piper, Jon K. Incrementally Rich Seeding Treatments in Tallgrass Prairie Restoration. *Ecological Restoration* 32, no. 4 (2014): 396–406. https://www.jstor.org/stable/43441687.
Ramer, Hannah, Kristen C. Nelson, Marla Spivak, Eric Watkins, JJames Wolfin, and MaryLynn Pulscher. "Exploring Park Visitor Perceptions of 'Flowering Bee Lawns' in Neighborhood Parks in Minneapolis, MN, US." *Landscape and Urban Planning* 189 (2019): 117–128.
Robbins, Paul. *Lawn People: How Grasses, Weeds, and Chemicals Make Us Who We Are*. Temple University Press, 2012.
Shober, Amy L., Geoffrey C. Denny, and Timothy K. Broschat. "Management of Fertilizers and Water for Ornamental Plants in Urban Landscapes: Current

Practices and Impacts on Water Resources in Florida." *HortTechnology* 20, no. 1 (2010): 94–106.

Steinberg, Ted. *American Green: The Obsessive Quest for the Perfect Lawn*. W. W. Norton, 2006.

Sturm, Alice, and Stephanie Frischie. "Mid-Atlantic Native Meadows: Guidelines for Planning, Preparation, Design, Installation, and Maintenance." Xerces Society for Invertebrate Conservation, 2020. https://xerces.org/sites/default/files/publications/19-052_MidAtlantic_Meadow_guidelines_web.pdf.

Takahashi, Masamichi, Kazuki Shibasaki, Eichiro Nakama, Moriyoshi Ishizuka, and Seiichi Ohta. "Use of Superabsorbent Polymer Resins in Forestry and Greening Fields." *Journal of Japan* 100, no. 6 (2018): 229–36. https://doi.org/10.4005/jjfs.100.229.

Verschoor, A., and N. Rethman. "The Possibilities of a Super Absorbent Polymer (Terrasorb) for Improved Establishment, Growth and Development of Oldman Saltbush (*Atriplex nummularia*) Seedlings." *South African Forestry Journal* 162 (1992): 39–42.

Volk, Michael I., Thomas S. Hoctor, Belinda B. Nettles, Richard Hilsenbeck, Francis E. Putz, and Jon Oetting. "Florida Land Use and Land Cover Change in the Past 100 Years." In *Florida's Climate: Changes, Variations & Impacts* edited by Eric P. Chassignet, James W. Jones, Vasubandhu Misra, and Jayantha Obeysekera. Florida Climate Institute, 2017.

Walker, E., and J. Banks. "Two-Stroke Engines in Landscape Maintenance: A Growing Public Health Problem." *Institute of Noise Control Engineering* 252 (2016): 99–105.

Weakley, Alan S., Derek B. Poindexter, Hannah C. Medford, Bruce A. Sorrie, Carol Ann McCormick, Edwin L. Bridges, Steve L. Orzell, Keith A. Bradley, Harvey E. Ballard Jr., Remington N. Burwell, Samuel L. Lockhart, and Alan R. Franck. "Studies in the Vascular Flora of the Southeastern United States: VI." *Journal of the Botanical Research Institute of Texas* 14, no. 2 (2020): 199–239. https://doi.org/10.17348/jbrit.v14.i2.1004.

Williams, M. J. "Native Plants for Coastal Dune Restoration: What, When, and How for Florida." US Department of Agriculture Natural Resources Conservation Service, 2007. https://www.nrcs.usda.gov/plantmaterials/flpmspu7474.pdf.

INDEX

Gage Daniel J. LaPierre, a lifelong Floridan, earned an MS and a PhD at the University of Florida (UF) studying restoration of southeastern pine savannas. During his graduate career he managed the UF Natural Area Teaching Laboratory for six years, conducting prescribed fire and restoration projects. Here, he founded the UF Native Plant Nursery Project with the aim of growing native plant materials for on-campus projects as well as for research, teaching, and extension activities. Starting in 2020 Gage began collecting written and anecdotal knowledge regarding the creation of novel "meadow" plant communities out of his strong interest in enhancing biodiversity in degraded greenspaces. Gage currently is a researcher at UF and lives in Gainesville, Florida.

Isabella Guttuso Browne is a photographer, landscape designer, and researcher with a passion for Florida's natural communities and an advocate for the use of native plants in urban environments. She is the Urban and Recreational Green Infrastructure Coordinator at UF's Center for Landscape Conservation Planning, where she conducts applied research on the role of green infrastructure in increasing resiliency, providing ecosystem services, and maximizing ecological function in Florida's communities. As a photojournalist for the Native Plant Horticulture Foundation and the Florida Association of Native Nurseries, she has captured the stories of nursery growers, designers, and homeowners who focus on native plant use to inspire and provide information about opportunities within the native plant industry. Isabella is a graduate of UF's Bachelor of Fine Arts and Master of Landscape Architecture Program. Isabella is a lifelong Floridian who currently resides in Central Florida with her husband and son, where they are actively converting their property to a native landscape.